2021 国际大学生建筑设计与数字建模竞赛作品集

丁 帅 主编

中国建材工业出版社

图书在版编目（CIP）数据

2021 国际大学生建筑设计与数字建模竞赛作品集 /
丁帅主编 . -- 北京：中国建材工业出版社，2022.3
ISBN 978-7-5160-3354-8

Ⅰ . ① 2… Ⅱ . ①丁… Ⅲ . ①建筑设计－作品集－世
界－现代 Ⅳ . ① TU206

中国版本图书馆 CIP 数据核字（2021）第 248149 号

内容简介

"一带一路"建筑类大学国际联盟秘书处组织编写的《2021 国际大学生建筑设计与数字建模竞赛作品集》
共收录竞赛征集的来自 13 个国家 34 所高校的 66 件作品，其中建筑设计方向作品 28 件，土木结构方向作品 21 件，
建筑与场景 3D 数字建模方向作品 17 件。

竞赛以"韧——唤起运河之魂"为主题，围绕"一带一路"运河古镇"修旧如旧"城市更新、可持续发展，
设立了建筑设计、土木结构、建筑与场景 3D 数字建模三个赛题方向，并附加运河古镇风情国际摄影大赛。

竞赛得到联合国教科文组织、国际摄影测量与遥感协会、中国测绘学会、G-Global 国际秘书处、中国丝路
集团有限公司、广联达科技股份有限公司、北京建大资产经营管理有限公司及"一带一路"建筑类大学国际联
盟成员高校的大力支持和参与，并由北京国际和平文化基金会、北京和苑博物馆共同协办。

竞赛以"专业性、权威性、创新性"为核心理念，以"以赛促教、以赛促学、以赛促培、以赛促融"和有
效提升联盟各成员高校学生的国际创新能力为宗旨，为"一带一路"建筑类大学搭建了科技创新与教育交流平
台，有效助力国际化创新型专业人才的培养，同时也为参赛学生提供了良好的实践机会和沟通融合的平台，引导、
促进了相关专业的教育教学改革。

本书可供广大工科院校师生以及建筑设计师、结构工程师、城市设计师研究参考。

2021 国际大学生建筑设计与数字建模竞赛作品集
2021 Guoji Daxuesheng Jianzhu Sheji yu Shuzi Jianmo Jingsai Zuopinji
丁　帅　主编

出版发行：中国建材工业出版社
地　　　址：北京市海淀区三里河路 1 号
邮政编码：100044
经　　销：全国各地新华书店
印　　刷：北京天恒嘉业印刷有限公司
开　　本：889mm×1194mm　1/16
印　　张：13.5
字　　数：240 千字
版　　次：2022 年 3 月第 1 版
印　　次：2022 年 3 月第 1 次
定　　价：198.00 元

本书编委会

主　　编　丁　帅

副 主 编　黄　鹤　黄　兴

参　　编　李　洋　刘　星　刘书青

　　　　　王　璇　田紫嫣　刘　璐

组织编写　"一带一路"建筑类大学国际联盟秘书处

虚里融渡

9:30A.M.

9:00A.M.

10:30A.M.

11:15A.M.

11:30A.M.

9:00-17:00

section 1-1

section 2-2

section 3-3

section 4-4

带电线圈
Live Coil

学校名称 University/College Name

中国·长春建筑学院

Changchun University of Architecture and Civil Engineering, China

指导教师 Supervisor (s)

张 蕾 ZHANG Lei　　　　　李建英 LI Jianying

参赛学生 Participant (s)

陈小雨 CHEN Xiaoyu　　　　项志坚 XIANG Zhijian

段 瑜 DUAN Yu　　　　　　胡隽玥 HU Junyue

陈晓玲 CHEN Xiaoling

简介 Description

在一个日趋多元和复杂化的社会中，如何让古运河重获新生，也应该是一种多元化的考量。

因此，对于运河沿线历史村落和码头的设计，我们跳出了传统的建筑设计范式，研究如何运用一些社会学、经济学和传播学的实践理论和方法，对业态、传播性、经济效益和文化价值等方面进行研究和规划，从而设计一种社会关系、一种传播符号、一种经济模式，而非单纯的建筑。

运河不仅可以用于运输，也可以成为传播媒介、社会关系、商业产品和文化载体。我们认为，唤醒运河的韧性之魂就是通过对这些属性的挖掘利用而实现。

In an increasingly diversified and complicated society, how to regain the ancient canal should also be a diversified consideration.

Therefore, for the design of historic villages and wharves along the canal, we jumped out of the traditional architectural design paradigm and studied how to apply the practical theories and methods of sociology, economics, and communication for research

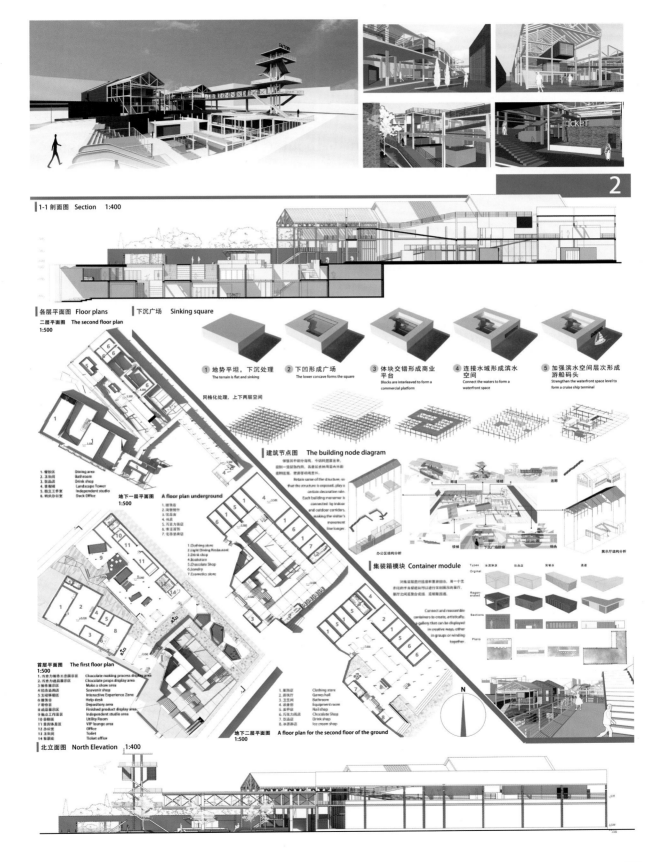

1-1 剖面图　Section　1:400

各层平面图　Floor plans　下沉广场　Sinking square

二层平面图　The second floor plan
1:500

① 地势平坦，下沉处理
The terrain is flat and sinking

② 下凹形成广场
The lower concave forms the square

③ 体块交错形成商业平台
Blocks are interleaved to form a commercial platform

④ 连接水域形成滨水空间
Connect the waters to form a waterfront space

⑤ 加强滨水空间层次形成游船码头
Strengthen the waterfront space level to form a cruise ship terminal

网格化处理，上下两层空间

建筑节点图　The building node diagram

地下一层平面图　A floor plan underground
1:500

集装箱模块　Container module

首层平面图　The first floor plan
1:500

地下二层平面图　A floor plan for the second floor of the ground
1:500

北立面图　North Elevation　1:400

姑苏渡口帆初落——基于古运河新生的客运码头设计

The Sails at the Gusu Ferry Are Just About to Fall—Based on the Design of the New Passenger Terminal of the Ancient Canal

学校名称 University/College Name

中国·河北建筑工程学院

Hebei University of Architecture, China

指导教师 Supervisor (s)

曹迎春 CAO Yingchun   班磊晶 BAN Leijing

参赛学生 Participant (s)

杨泽泉 YANG Zequan   刘延冰 LIU Yanbing

李宗阔 LI Zongkuo   陈晓玲 CHEN Xiaoling

高一民 GAO Yimin

简介 Description

"姑苏渡口帆初落，渔浦山头日未欹"，苏州园林作为苏州的城市名片，同样也承载着历史文化名城之精神。本次方案设计以客运码头为中心，以古典园林的设计意象为纽带，连接京杭大运河与苏州古城，在古运河"苏州段"融入地域文化特色，最终形成集客运、展览和景观三大功能于一体的城市古运河客运码头。作为对城市革新的一次全新的建筑探索，设计方案为城市注入了新鲜活力，从而带动城市运河沿岸"新陈代谢"，使古运河重获新生。

Suzhou gardens, as the name card of the city, also carry the spirit of this historical and cultural city. Centering on the passenger terminal, the design proposal adopts the symbol of the classical garden as a link between the Beijing-Hangzhou Grand Canal and the ancient city of Suzhou, integrating the Suzhou reach of the ancient canal into its regional culture. Eventually, an urban ancient canal passenger terminal is formed, which combines three functions, i.e. passenger terminal, exhibition, and sightseeing. As a new architectural effort to explore urban renewal, the proposal injects fresh vitality into the city, thus driving the "metabolism" along the urban canal and reinvigorating the ancient canal.

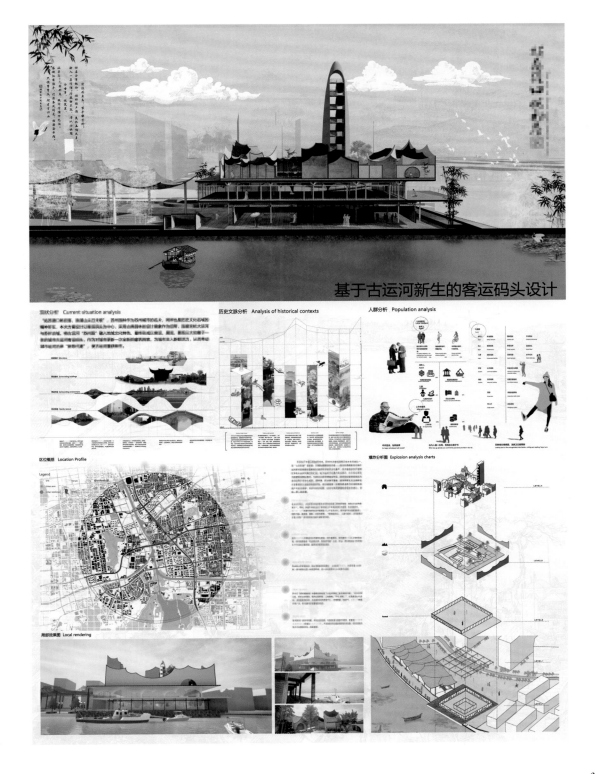

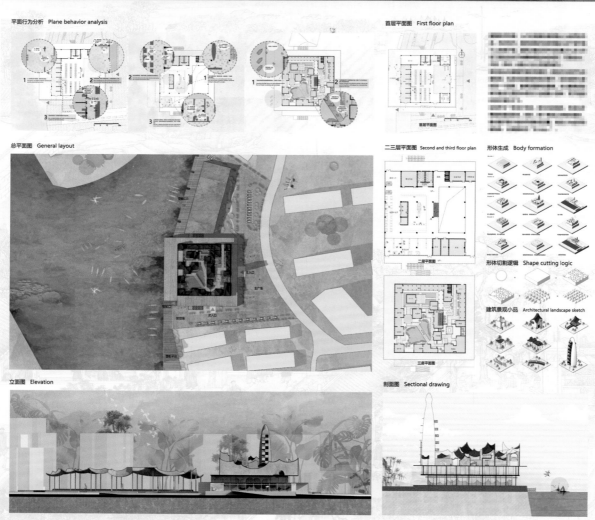

平面行为分析 Plane behavior analysis

首层平面图 First floor plan

总平面图 General layout

二三层平面图 Second and third floor plan

形体生成 Body formation

形体切割逻辑 Shape cutting logic

建筑景观小品 Architectural landscape sketch

立面图 Elevation

剖面图 Sectional drawing

中国长江与大运河交汇处的世业洲旅游港口项目

Tourist Port Project on the Island Shiye at the Intersection of the Yangtze River and the Grand Canal of China

学校名称 University/College Name

亚美尼亚·亚美尼亚国立建筑与建设大学

National University of Architecture and Construction of Armenia, Armenia

指导教师 Supervisor (s)

Anna Yengoyan   Anahit Vardanyan

参赛学生 Participant (s)

Martin Martikyan   Melanya Hakobyan

Viktorya Yeryomina   Yana Arakelyan

简介 Description

京杭大运河是世界上历史最悠久的水工建筑物之一，目前仍在运行当中，它连通着中国南北，东西方则由长江连接。这两条主要交通路线的交汇处被称为"黄金水道"。在两大旅游城市镇江和扬州之间，坐落着一个与世业镇同名的世业洲。该岛屿地理位置优越，毗邻主要交通路线，根据国家发展规划，拟在岛屿东端修建河港和泊位。此外，该项目还将对全境进行完善与开发，形成体育、娱乐、休闲等功能区。岛上现有的酒店建筑群将借助桥梁融入整体布局，该桥梁使用著名的拱桥结构。该构想将使这一区域兼具舒适性、竞争力，更重要的是，它将备受欢迎且紧跟时代潮流。

The Grand Canal of China is one of the oldest and the currently operating hydraulic structures in the world, which is connecting the south of the country with the north, and the Yangtze River, connecting the east and the west. At the crossroads of these two major transportation routes, named Golden Route, between the largest tourist cities Zhenjiang and Yangzhou, is located the Shiye island, with a city of the same name. Taking into account the favorable location of the island, the proximity to the main transport routes, as well as the state program for the development, it is proposed to place a river port and berths on the eastern tip of the island. In addition, this design also aims to improve and develop the entire territory, form sports, entertainment and leisure zones. The existing hotel complex on the island is integrated into the overall concept with the help of a bridge, the prototype of which is the famous lunar bridge. The proposed concept makes it possible to form a comfortable, competitive, and most importantly, in-demand and constantly modernizing architectural environment for such a significant, in our opinion, territory.

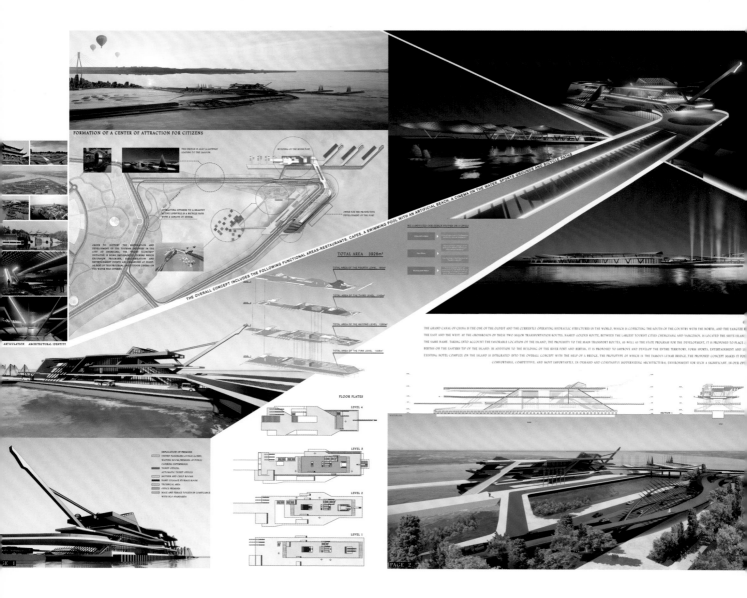

岛 客
D.O.C.K.

学校名称 University/College Name
中国·天津城建大学
Tianjin Chengjian University, China

指导教师 Supervisor (s)
万 达 WAN Da   Przemysław Kowalski

参赛学生 Participant (s)
王秉仁 WANG Bingren   高 亮 GAO Liang
秦智轩 QIN Zhixuan   丁雪娣 DING Xuedi

简介 Description

Pł awniowice 目前是一个稍显没落的波兰村庄，位于宫殿建筑群中，其核心功能是宗教中心。该方案旨在运河和湖泊之间的狭长土地上建造一个兼具客运港口（于运河上）和划船俱乐部（于湖泊中）功能的文化综合体，以激活当地的旅游需求，并为区域和当地社区服务。

我们提取了当地传统的坡屋顶元素，并结合水的波浪形式生成了屋面曲线；从城堡的木质龙骨中提取元素，形成方案中的木质结构；同时，应用多种主被动生态技术，以期尽可能地减少对环境的影响。

Pł awniowice is a rather sleepy village in Poland, mainly functioning as a religious center located in the palace complex. We proposed a cultural complex with a passenger port (on the canal) and a rowing club (in the lake) on the narrow land between the canal and the lake, to stimulate the local tourist demand and provide regional and local community services.

In terms of form generation, we extracted the roofline of the traditional local sloping roof and combined it with the waveform of the water to generate the roof curve. In terms of building materials, we extracted elements from the wooden keel of the castle roof to create the wooden structure of the scheme. Meanwhile, we also applied both active and passive ecological technologies to make the environmental impact of the building as low as possible.

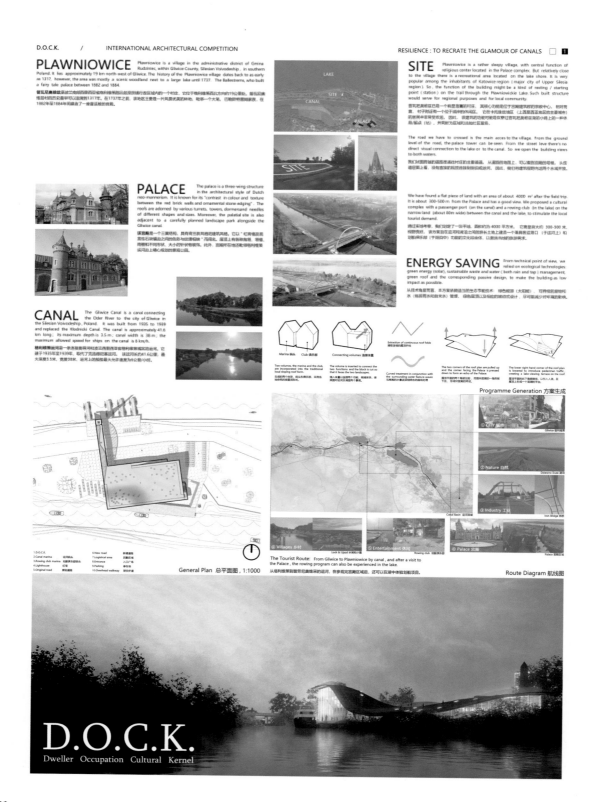

连舟浮桥·点线共生
——运河上衰败码头的活力再生与生态恢复

Pontoon Bridge, Coexistence along the Line—Vigorous Regeneration and Ecological Restoration of Decaying Canal Wharf

学校名称 University/College Name

中国·安徽建筑大学
Anhui Jianzhu University, China

指导教师 Supervisor (s)

桂汪洋 GUI Wangyang   周庆华 ZHOU Qinghua

参赛学生 Participant (s)

李成成 LI Chengcheng   吴扬扬 WU Yangyang
吴静文 WU Jingwen   张辉辉 ZHANG Huihui
卢　泉 LU Quan

简介 Description

随着交通方式的发展和进步，运河的航运功能逐渐减弱，导致运河沿线活力丧失，城市衰败，生态问题严重。为了焕发运河的新生，本设计提出"连舟浮桥，点线共生"的概念。连舟浮桥为形式，沿线城市的共生为目的。以隋唐大运河临涣码头为典型案例，建立一个"可移动的聚落"，该聚落涵盖生产、生活、生态保护与休闲交流等功能，为当地居民提供具有活力的公共空间。由点到线，以"可移动的聚落"作为触媒，在古运河航线的不同码头以不同的形态绽放，逐步促进运河沿岸各城镇的联系与发展，为运河发展注入新的活力。

With the development and progress of transportation, the shipping function of the canal gradually weakened, resulting in the loss of vitality along the canal, urban decline, and serious

2021 国际大学生建筑设计与数字建模竞赛作品集
2021 International Student Competition on Architectural Design and Digital Modelling Work Collection

ecological problems. In order to rejuvenate the canal, the design puts forward the concept of "pontoon bridge, coexistence along the line". The pontoon bridge is designed as a way to achieve the coexistence of cities along the canal. Taking the Linhuan wharf of the Grand Canal in the Sui and Tang Dynasties as an example, a "movable settlement" was established, which covered the functions of production, living, ecological preservation, leisure, and communication, and provided a vibrant public space for local residents. From point to line, "movable settlements" are used as a catalyst to blossom in different forms at different docks of the ancient canal route, gradually promoting the connection and development of towns along the canal, and injecting new vitality into canal development.

水畔四方街——文化基因视角下的沙溪古镇码头设计

Sifang Square by the Riverside—Wharf Design of Shaxi Ancient Town from the Perspective of Cultural Genes

学校名称 University/College Name

中国·重庆交通大学
Chongqing Jiaotong University, China

指导教师 Supervisor (s)

温 泉 WEN Quan   徐 辉 XU Hui

参赛学生 Participant (s)

邱雅雄 QIU Yaxiong   孙思可 SUN Sike

郭书伶 GUO Shuling   张珏澜 ZHANG Juelan

韦乔焯 WEI Qiaozhuo

简介 Description

本设计选址于世界建筑文物保护基金会（WMF）保护地，国家历史文化名镇云南沙溪的黑惠江畔。设计提取当地四方街、登山廊、老戏台、夯土墙……这个茶马古道上唯一幸存的集市的丰富历史文化基因，从乡土材料、构筑方法、公共空间、民俗文化几个方面提炼了丰富的景观基因，并通过拼合、镶嵌、整合的方法进行传承。通过延续乡土材料肌理，传承四方街空间结构，以及重塑古镇水滨生活场景，希望码头能带领乡民和游客们重回黑惠江畔，让这里的声音传到全世界。

The site is located at the Heihui River bank of Shaxi, a national historical and cultural town in Yunnan Province as well as a protected area under World Monuments Fund (WMF). The design extracts historical and cultural genes of Sifang Square, mountain climbing gallery, old stage, rammed earth wall, which composes the only surviving market on the ancient tea horse road. It extracts landscape genes from the aspects of building materials, construction methods, public space and folk culture, and inherits them through the methods of combination, inlay and integration. By continuing the texture of local materials, inheriting the spatial structure of Sifang Square, and reshaping the waterfront life scene of the ancient town, we hope that the wharf will bring the villagers and tourists back to the Heihui River again, and let the horseshoe bell spread to the world.

水畔四方街
——文化基因视角下的沙溪古镇码头设计

元素来源

使用材料

青瓦　木材　石　玻璃　夯土

景观分析

爆炸图分析

空间分析

通风采光

视线分析

1-1剖面图1：200　　2-2剖面图1：200　　临江阁展建筑北立面图1：200

前码今生

Former Wharf in Present Life

学校名称 University/College Name

中国·吉林建筑大学
Jilin Jianzhu University, China

指导教师 Supervisor (s)

宋义坤 SONG Yikun   安　宁 AN Ning

参赛学生 Participant (s)

孙文昊 SUN Wenhao   吴哲涵 WU Zhehan

戴沈周 DAI Shenzhou   郭　璐 GUO Lu

陈嘉怡 CHEN Jiayi

简介 Description

码头不仅是古往今来人们生存的途径和场所，也是连接生产、生活与运河的纽带，记忆着数千年来人们对美好生活的向往和不懈的奋斗，以及运河发生的故事。本设计依托新的规划，旨在发掘并重新阐释码头的历史记忆。

Wharf is not only the means and place for people to survive, but also the link between production, life, and the canal. It records people's yearning for a better life and hard struggle for thousands of years, and stories taking place along the canal. The purpose of this design is to explore and reinterpret the historical memory of the wharf against new formats.

Hongbo Wharf in Nanjing

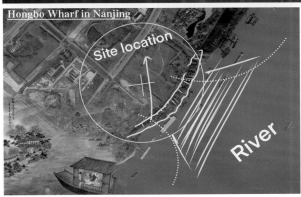

Site location

River

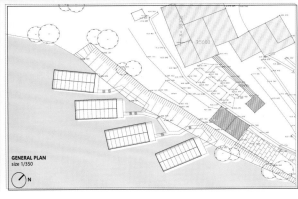

GENERAL PLAN
size 1/350

N

Design intent

Possibility study

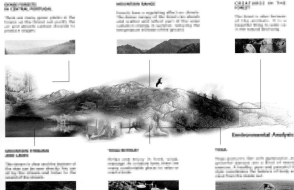

Environmental Analysis

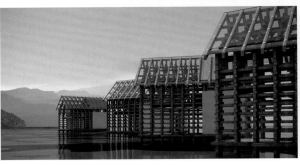

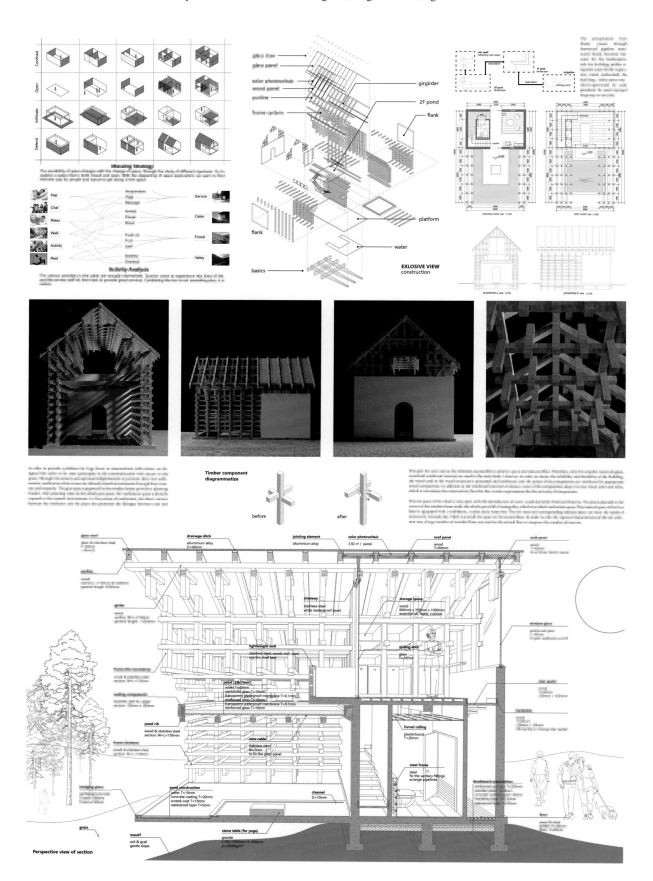

泽波斯船长皮划艇公园

Kapitein Zeppospark

学校名称 University/College Name
法国·犹尼亚高等工程师学院
Junia Graduate School of Engineering, France

指导教师 Supervisor (s)
Vedrana IKALOVIC

参赛学生 Participant (s)

Nomindari ENKH–AMGALAN   Rida ASSAAD

HU Wentao   Fanny GOMBET–HOMEGA

简介 Description

在第二次世界大战前，位于根特运河的 Houtdok 码头主要用于本地及外来船只装载和储存木材及其他货物，此后却被忽视了，直到去年，一项城市更新项目提议通过 Kapitein Zeppospark 让码头焕发生机，并为海滩增添新的乐趣。我们以此为先机，对该项目进行了扩展，从而加强该地的吸引力和实用性，以重现其昔日的热闹景象。我们弥补了周边社区（亚洲美食餐厅、图书馆、艺术展、健身房、皮划艇中心）和城市（植物园）缺失的服务和活动，并进一步与附近的酒吧和社区中心（年轻人有机会在此练习烹饪，观看电影和戏剧，体验木工和文化艺术，享受美食、音乐会、派对和饮料）建立联系。通过以上合作，我们为艺术展及此处的其他地段带去新的机会、活动和联系。最后，为了提高市中心的吸引力，我们建造了一个皮划艇中心，让人们能够体验环城皮划艇运动。

Following its usefulness in uploading and stocking timber wood and other goods from local and outside ships before World War Ⅱ, the "Houtdok" dock in the canal of Ghent became neglected by the city until last year, when an urban renewal project was proposed to give the dock a new greener look with the "Kapitein

Zeppospark", and a new fun aspect with the beach. We took this opportunity as a head start to expand the project and make the site more attractive and more usable in order to regain its busyness like old times. We added services and activities which are missing from the surroundings (Asian Food Restaurant, library, art exhibition, gym, kayaking center) and from the city (Botanic Garden). We expanded more by making a connection with a bar and a community center (in which youths have the opportunity to practice cooking, see a film, go to a theater, do woodworking, feel the cultural art and enjoy food, concerts, parties, and drinks) by our side, from which we will cooperate to increase opportunities, events, and connectivity in our art exhibition and the rest of the site). Finally, in order to add more attractivity of the city center, we constructed a Kayak center, where people can practice the kayak sport around the city.

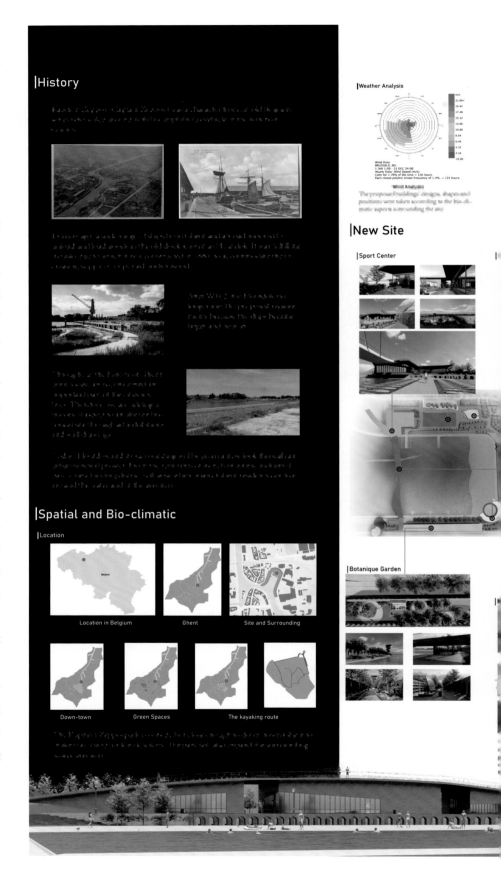

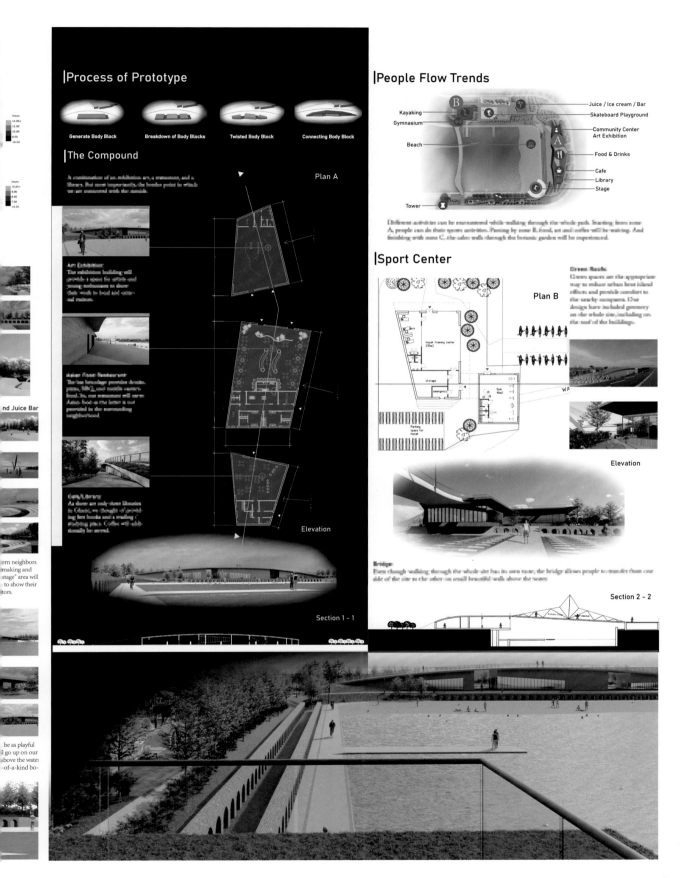

Process of Prototype

Generate Body Block | Breakdown of Body Blocks | Twisted Body Block | Connecting Body Block

The Compound

Plan A

Section 1 - 1

People Flow Trends

Kayaking
Gymnasium
Beach
Tower

Juice / Ice cream / Bar
Skateboard Playground
Community Center
Art Exhibition
Food & Drinks
Cafe
Library
Stage

Sport Center

Plan B

kayak training center 225m2

storage

emergency

Gym 90m2

Parking space for kayak

Elevation

Section 2 - 2

巴拿马运河甘博亚镇客运港设计

The Passenger Port Design in the town of Gamboa, Panama Canal

学校名称 University/College Name

亚美尼亚·亚美尼亚国立建筑与建设大学
National University of Architecture and Construction of Armenia

指导教师 Supervisor (s)

Zaruhi Mamyan   Erzas Hovhannisyan

参赛学生 Participant (s)

Angela Sargsyan   Hayk Ziraqyan

Rita Deek

简介 Description

设计选址位于巴拿马甘博亚镇，该地区以热带雨林而闻名。我们的目标是鼓励当地发展生态旅游，并为当地注入新的活力。

该地区有一座古老的灯塔，修建于巴拿马运河建设期间，遗憾的是现在已经被废弃。本项目的主要目标之一便是对该灯塔进行修复。沿海岸线有一条铁路，考虑到镇上没有火车站，我们决定为该建筑群增加火车站这项额外功能。

因此，本项目具有三个主要功能：火车站、客运港和运河博物馆。建筑群由两栋建筑组成，通过隧道和桥梁连接起来。

正门建筑设有博物馆，其中一部分直接坐落于灯塔内部，形成了一个旅游景点。我们试图为新旧建筑之间建立联系，该建筑群被命名为"光之屋"，第二栋建筑将用作港口和火车站。

旅客到达该地后，过桥即可进入博物馆，在那里他们将欣赏到甘博亚镇的自然风光，了解当地的历史。

解　锁

Unlocked

学校名称 University/College Name

英国·东伦敦大学

University of East London, the UK

指导教师 Supervisor (s)

Fulvio Wirz   Heba Elsharkawy

参赛学生 Participant (s)

Simone Pamio   Viktor Telecky

Aleksandra Ewa Hoffmann   Ahmad Wahid Feroz

Mahabub Tuhin Alam

简介 Description

名为"解锁"的重建方案对象是伦敦的三条运河，即摄政运河、莱姆豪斯运河、赫特福德运河，此外还包括利河。

历史上，在工业革命期间，这几条运河交通繁忙，相互连接，从而加快了原材料的运输。船闸的使用便利了航行，解决了运河间巨大的高度差问题。

如今，这几条运河几乎已不再发挥其原始作用，而是专供中型船只使用，市民只能在运河周围走动。小型船只使用者则不愿意使用船闸，因为那将会浪费大量时间和水资源，极其不实用。因此，他们只能在水上进行短距离航行。

"解锁"设计提出了一种运河航行的创新方案，用传送带解决了船闸所带来的困难。为了促进平稳过渡，本项目引入了一款名为"LockNess"的新船，该船型可自动驾驶，并可在返回港口期间对运河进行清洁。

2021 国际大学生建筑设计与数字建模竞赛作品集
2021 International Student Competition on Architectural Design and Digital Modelling Work Collection

麦尔安德港口作为主要网络枢纽，凭借其特性和形式带动着小型辅助站点的发展。如今，市民可以自由地航行或停歇，并探索城市各地。

该方案旨在将城市结构相互连接，并增强水资源的可获得性。

主港口将围绕着具有历史意义的船闸打造一个区域，帮助人们提高对伦敦独特的工业遗产的认识。

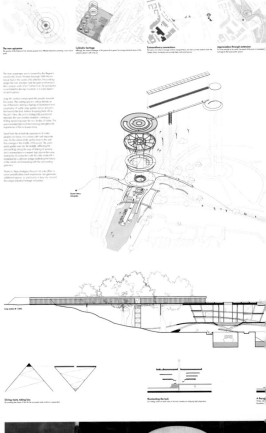

"Unlocked" is a proposal for the redevelopment of 3 canals in London: Regents Canal, Limehouse Canal and Hetford Union Canal, together with River Lea.

Historically the canals were heavily trafficked and joined together to speed the movement of raw materials during the Industrial Revolution. The use of the locks facilitated navigation and allowed to overcome large height differences throughout the canals.

Today, the canals are virtually unused for their original purpose, used exclusively by medium-sized boats. Citizens can only experience the canals by walking around them. Small boats users are reluctant to use the locks as they pose a significant waste of time and water, being highly impractical. The water experience is, therefore, limited to short distances.

"Unlocked" design proposes an innovative solution in navigating the canals, solving the difficulty posed by the locks with conveyor belts. A smooth transition is reinforced by the "LockNess": a new boat capable of self-driving and cleaning the canal during its return journey back to the port.

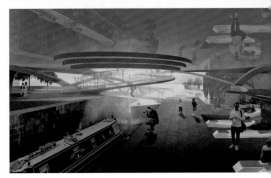

Mile End Port becomes the main network hub sharing its qualities and lending its form to the smaller subsidiary stops. Citizens now have the freedom to continue sailing or stop and explore various parts of the city.

The proposal aims to interconnect the urban fabric and facilitate access to the water.

The main port strives to create a stage around the historical lock, helping to raise awareness about London's unique industrial heritage

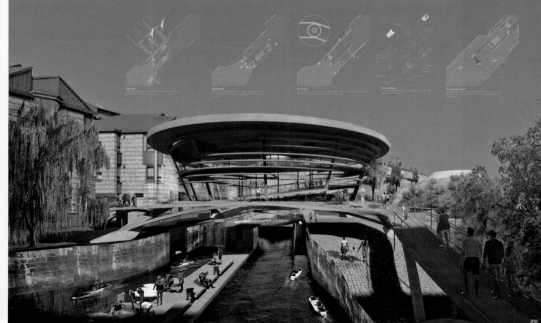

奇妙的 Begej 运河

Absurd Begej Channel

学校名称 University/College Name

塞尔维亚·诺维萨德大学
University of Novi Sad, Serbia

指导教师 Supervisor (s)

Ivana Miškeljin   Marko Jovanović

参赛学生 Participant (s)

Staša Zeković   Iva Pejčić

Anastasija Radovanović   Marko Mihajlović

Mladen Kesegić

简介 Description

本项目选址于兹雷尼亚宁市（塞尔维亚）Begej 运河，此处受溪流冲刷形成一个岛屿。该岛屿形态独立，兹雷尼亚宁城区的历史及自然文化遗产丰富，这里有法院、检察院、监狱、旧工厂、教堂和多个公园，因而被称为"小美国"。该地区一个最引人注目的特点是，这里拥有世界上唯一一座横跨裸露地面的桥梁，在当地被称为"旱地之桥"或简称为"旱桥"。为了强化宣传这一当地特色，本项目选题奇妙而鲜明，建造了一座纵向环岛"水"桥，而不是通常的横向桥梁，也不同于兹雷尼亚宁现有的"旱"桥。此外，由于全球变暖，旱桥下的干燥地面重新浮出了水面，因此在船港设计中，我们着重于通过反射实现通道的视觉延续，并确保其狭窄的水道可供游船通行。

Our project is located on Begej channel in the city of Zrenjanin (Serbia), on the spot where the stream of the channel creates an island. This island is called "Little America" because of its isolated morphology and is an interesting urban area of Zrenjanin filled with cultural and natural heritage, such as the buildings of the courthouse, the prosecution office, the jail, old factories, churches and many parks. One of its most striking features is the only bridge in the world that spans bare ground and therefore is called locally "The Bridge on Dry Land" or just "Dry Bridge". In order to underline and promote this local phenomenon, our absurd and contrast project is creating a longitudinal "water" bridge which wraps the island, instead of a usual transversal, and is also different from the existing "Dry Bridge". This dry ground underneath the "Dry Bridge" resurfaced, inter alia, due to global warming. The aim of the design is to create visual continuation of the channel by using reflection and assuring its function of waterway slits for tourist boats tours as well as a boat port.

腾飞！伊斯梅利亚！

Up! Ismailia!

学校名称 University/College Name

中国·北京建筑大学
Beijing University of Civil Engineering and Architecture, China

指导教师 Supervisor (s)

刘平浩 LIU Pinghao   李 欣 LI Xin

参赛学生 Participant (s)

史祚政 SHI Zuozheng   崔 涵 CUI Han

高 源 GAO Yuan   梁家澍 LIANG Jiashu

李 想 LI Xiang

简介 Description

本项目位于苏伊士运河上的伊斯梅利亚古镇，通过全新的游轮码头设计，力求在功能层面打通游轮、游船停靠的基本使用，进而在精神层面，为这样一个立足于苏伊士运河的数百年古镇注入新的活力。

设计依托古镇的历史轴线，面向苏伊士运河入海口，建造物拔地而起，在伸向大海的同时，向天空"腾飞"，完成了历史与未来的对话，成为运河的纪念碑，也必将成为城市的全新地标，标志着城市文化发展的腾飞。

源自运河的文化底蕴，指引运河发展的光明未来。

This design is about the ancient town of Ismailia by the Suez Canal. Through the design of the new cruise terminal, it seeks to provide a feasible way to berth cruises and ships, and more importantly, inject new vitality into the century-old town.

The design builds along the central axis of the ancient town. At the estuary, the construction extends towards the sea and rises into the sky

2021 国际大学生建筑设计与数字建模竞赛作品集
2021 International Student Competition on Architectural Design and Digital Modelling Work Collection

from the ground. As a monument to the canal, it links history and the future and will become a new landmark of the city, heralding the city's cultural take-off.

The design draws inspiration from the history of the canal and ushers in its bright future.

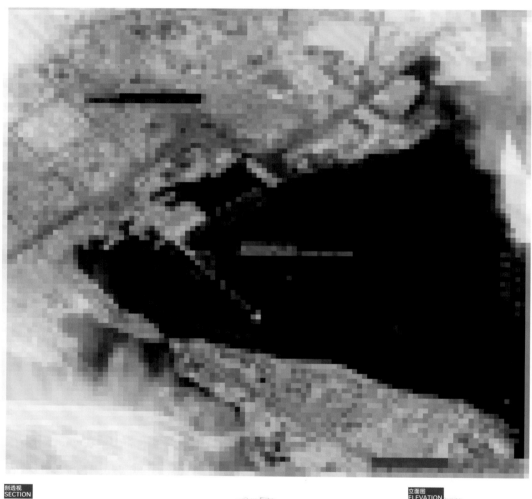

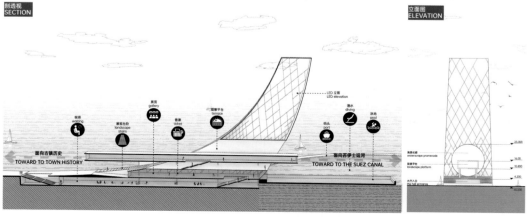

莫斯科运河是"五海之港",五海是指白海、波罗的海、里海、亚速海和黑海。主要货运码头是莫斯科北河码头。

运河项目的核心是一个径向系统,该系统由码头和通向灯塔、停车场的桥梁聚合而成。港口和游艇俱乐部的建筑都位于形似罗盘指针的屋顶下。该项目主要以圆形呈现,状如地球,近似莫斯科的交通系统,给人以旅行之感。

The site is in the microdistrict Hlebnikovo of Dolgoprudny, Moscow, at the intersection of the Dmitrov highway and the Moscow Canal as well as by the reservoir of Klyazma.

The port at the intersection is beneficial as a place of transfer hub. By the Dmitrov highway people can reach the international Sheremetyevo Airoport, the Savelovsky Railway station, Dmitrov and Dubna. Also the port is near to the Moscow Central Diameter D1.

The location of the port is a good decision. The town of Dolgoprudny has a rich history which is connected with shipbuilding. Here is located a former Hlebnikov machine-building and ship-repairing plant/factory which was transformed into the Moscow Yacht Port. It turned out that the area of the Canal in the territory of Dolgoprudny became a historical place of yacht clubs' location, and it was reflected in our project. Apart from that, the infrastructure of the town requires creating a new place of people's attraction, free recreational place for rest and communication. At the moment most of the riverside area of the Moscow Canal in the territory of the town is a private property. The transport congestion of the Canal is quite serious. By the Canal pass cruise routes in various directions, people make cargo transportations here. The project of passengers' port includes the building of the river station (square) and the building of the yacht club (square), the big well-maintained parking zone, the public beach, the system of docks/piers, the lighthouse - a museum of shipbuilding.

The Moscow Canal is a "port of five seas", the White Sea, the Baltic Sea, the Caspian Sea, the Sea of Azov and the Black Sea. The main cargo terminal is the North River Port of the capital city.

The core of the project is a radial system which is formed/created by forms of docks/piers and the bridge which leads to the lighthouse and the parking zone. The buildings of the port and the yacht club are located under the same roof that resembles a form of compass needles. The leitmotif of the project is the circle form which resembles a form of the Earth, the transport system of Moscow, and it instills a spirit/a sense of journey.

惠享运河时代，重塑千年古韵
——沧州大运河客运码头设计

Reshaping the Ancient Charm
—Design of Cangzhou Grand Canal Passenger Terminal

学校名称 University/College Name

中国 · 天津交通职业学院

Tianjin Transportation Technical College, China

指导教师 Supervisor (s)

纪欢乐 JI Huanle   张宏涛 ZHANG Hongtao

参赛学生 Participant (s)

张荣涵 ZHANG Ronghan   陈　涛 CHEN Tao

孙明杰 SUN Mingjie   马大器 MA Daqi

张　宇 ZHANG Yu

简介 Description

　　京杭大运河沧州段位于沧州市中部，全长 253 公里，沧州是京杭大运河流经的 20 个城市中里程最长的城市。它保持了历史河道的原真形态，独具北方特色。

　　本设计范围为大运河沧州段市区段。设计范围内总面积 3800 余亩，其中绿地面积 1000 余亩，水系面积 800 余亩，共设置学校 1 所，主题公园 1 个，客运码头 1 个，货运码头 1 个，文化长廊 1 个。

Cangzhou section of Beijing-Hangzhou Grand Canal is located in the middle of Cangzhou City, with a total length of 253 kilometers. It is the city with the longest mileage among the 20 cities that

the Beijing-Hangzhou Grand Canal flows through. It maintains original shape of the ancient river and boasts unique of northern characteristics.

　　The scope of the design is Cangzhou urban

section of the Grand Canal. The design area covers total of more than 3,800 mu (1 hectare=15 mu), including more than 1,000 mu of green land and more than 800 mu of water system. There is a school, a theme park, a passenger terminal, a cargo terminal and a cultural corridor.

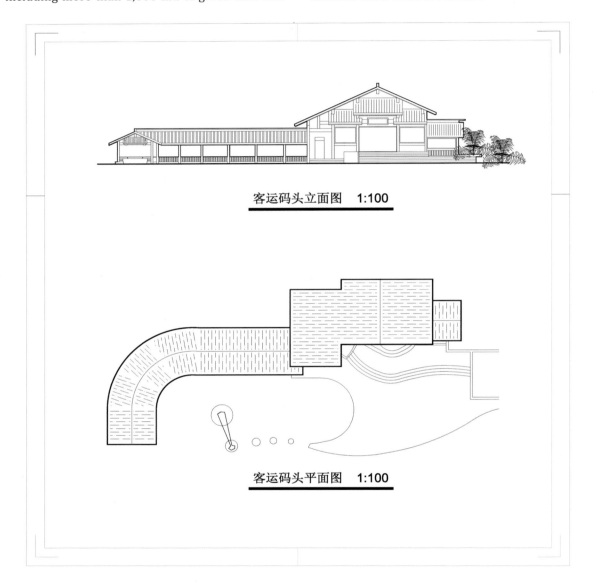

客运码头立面图　1:100

客运码头平面图　1:100

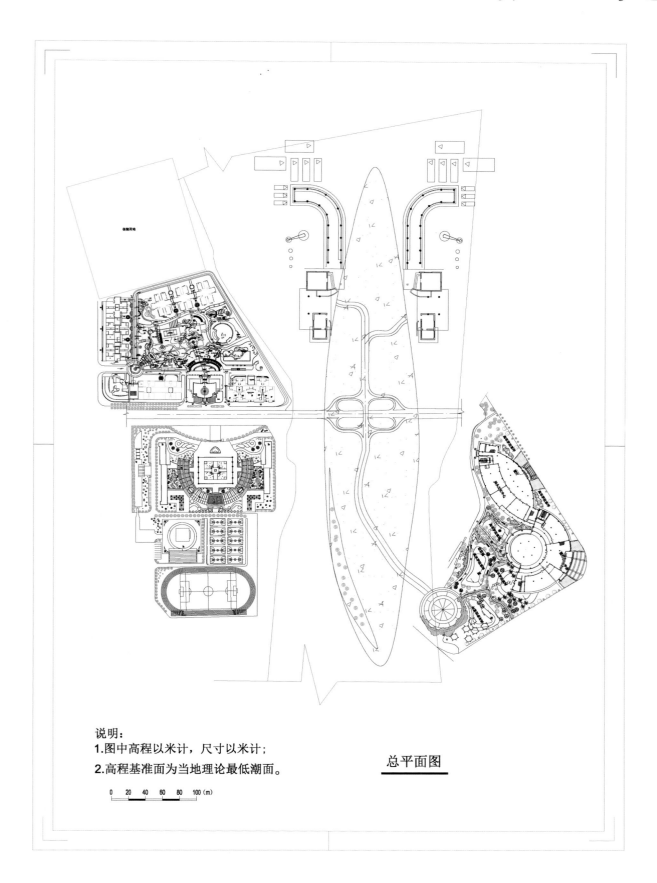

说明：
1.图中高程以米计，尺寸以米计；
2.高程基准面为当地理论最低潮面。

总平面图

0 20 40 60 80 100 (m)

运河，与城市同生共长

ProDe

学校名称 University/College Name

黑山共和国·下格理查大学

University of Donja Gorica, Montenegro

指导教师 Supervisor (s)

Bojana Sternisa   Semso Kalac

参赛学生 Participant (s)

Aldijana Hodzic   Tamara Mracevic

Sara Djikanovic   Arman Pepeljak

简介 Description

在城市设计中，水元素至关重要，因为它形状多变，启发无穷。运河是货物运输和游客旅行的关键渠道，也是文化交流的重要方式。运河不是一条线，而是一个网络，有各种功能和扩建工程，连接着港口和城市活动。它将成为城市的新核心，并设有公园和步行街。这个重建项目将包括不同类型的滨水开发项目，所以我们下一步将滨水地区作为现有城市建筑结构的一部分进行开发，呼应已明确的历史特征。我们设计时将该项目的"多功能性"放在首要位置，为人们提供各种社交、商业和文化用途。为了使人们获得更加愉快的体验，我们优先考虑的是通道设计，这样景色才会更有吸引力。灵活性同样是我们设计理念的关键之处，以此可在中国实现可持续发展。指标和关键因素包括：交通升级；灵活的环境；公共空间和绿地；增加滨水地区通道；多样的活动；重塑土地利用格局和景观。我们看到了这个城市的潜力，通过这个项目，我们有机会表达自己，展示我们如何看待中国及其在各方面的潜力。

Water as a design element in an urban area is significantly crucial, it fulfils many angles and is inspirational. Watching canals as important channels for good and people to travel inspired us to see them as cultural exchanges. The canal won't be considered as a line, it is imagined as a network of places, functions, additions and hinges between the port and urban activities. Our design will be new core of city with parks and promenades. This regeneration will include different types or typologies of waterfront developments, and waterfronts as part of the existing urban fabric will give an answer to history identity of already defined character. Mixed use is priority, it will offer people different social, commercial and cultural uses. Access must

be prioritized so views will be attractive in order to allow people more enjoyable time. Flexibility is a key for our idea, which aims sustainable growth in China. Indicators and key factors indude upgraded transportation, flexible environment, public spaces and green spaces, increasing waterfront access, different activities, reshaping land use pattern and landscape. We see potential in this part of the city as an opportunity to express ourselves, through this project We show how we see China and all the potential it has.

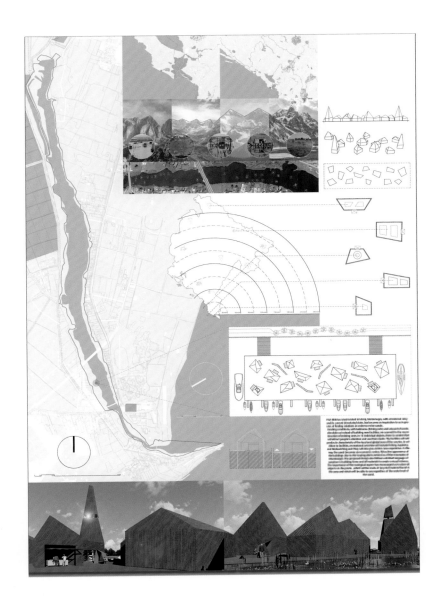

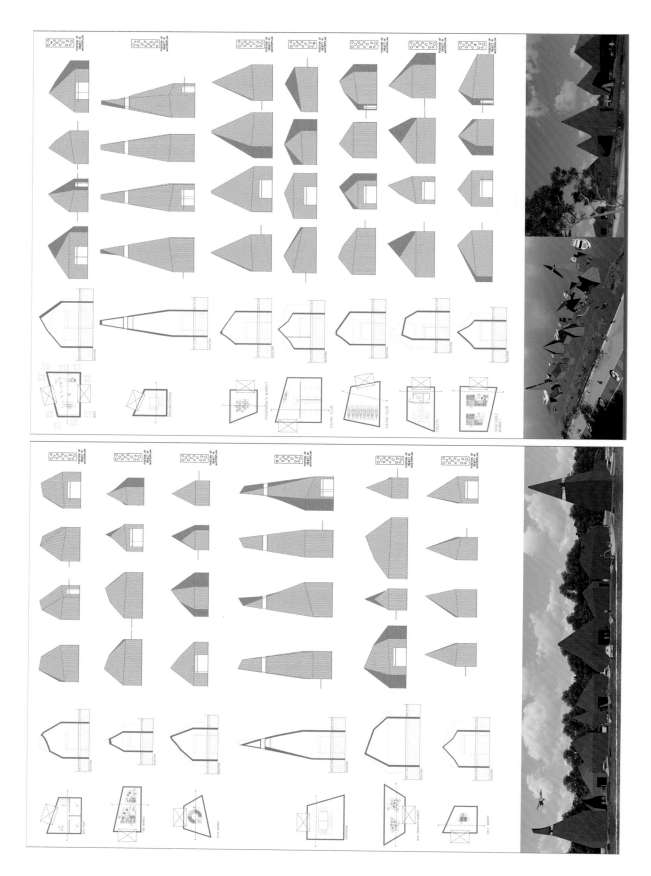

土木结构方向
Category B: Structural Design

舟楫万里

The Boat Sails for Thousands of Miles

学校名称 University/College Name

中国·北京建筑大学
Beijing University of Civil Engineering and Architecture, China

指导教师 Supervisor (s)

焦驰宇 JIAO Chiyu

参赛学生 Participant (s)

赵宇昂 ZHAO Yu'ang   谭希学 TAN Xixue

张云开 ZHANG Yunkai   王铭鑫 WANG Mingxin

全振鹏 QUAN Zhenpeng

简介 Description

本桥梁采用斜拉结构，桥塔上的花纹描绘着京杭大运河昔日的繁荣景象以及大运河的航线，表示着通州运河文化的起源。桥塔主体参考古代商船的船帆造型，两侧的人行道条带宛如运河上的波浪。整体看去，桥梁犹如一艘巨型商船行驶在波涛汹涌的北运河之上，描绘出一幅"蔽空千帆，赫然繁华之琳琅"的壮丽景色。寓意中国这艘巨轮在为实现中华民族伟大复兴的中国梦的航线上，不断高歌猛进，乘长风破万里浪！

The bridge adopts a cable-stayed structure, and the pattern on the tower depicts the prosperous scene and route of the Beijing-Hangzhou Grand Canal. It indicates the origin of Tongzhou canal culture. The main body of the bridge tower refers to the sail shape of ancient merchant ships, and the sidewalks on both sides like waves of the canal. On the whole, the bridge likes a giant merchant ship sailing on the turbulent North Canal, depicting a magnificent scene of "thousands of sails in the air, splendidly prosperous". It means that China, the great ship, is making great progress on the route of realizing the Chinese dream of the great rejuvenation of the Chinese nation, riding the long wind and breaking the waves!

莫斯科运河区斜拉公共空间桥梁设计

Cable-Stayed Public Space Bridge Design in Moscow Canal Area

学校名称 University/College Name

俄罗斯·莫斯科国立建筑大学
Moscow State University of Civil Engineering, Russia

指导教师 Supervisor (s)

Magomedov Marat   Saltykov Ivan

参赛学生 Participant (s)

Papikyan Karine   Zenkina Alisa

Zhuravleva Daria

简介 Description

使用拱门和拉索作为结构元素的想法诞生于项目开发早期，本项目的设计概念源于与河流的联系，取河鱼的骨架作为建筑物的寓意。此外，通过对该概念的详细研究，我们决定将拱形结构简化为桥梁的中心语义轴。因此拱门并非平行通过路基，而是斜向通过路基，打造了一个更具动态的结构，同时也更加美观。在设计道路时，我们立刻想到要为公共区域而不是交通提供大部分空间，但也并没有忘记现有的交通路线。因此，考虑到 Serebryany Bor 地区距离很近，适合个人交通工具，我们决定只为公共交通工具（电动巴士）设立双车道，两条车道宽度为 7 米，自行车道宽度为 2.5 米，带景观区的人行道宽度为 9 米。

The idea to use arches and cables as structural elements came at the early stages of development. In the early stages, the concept rested on the connection with the river, so the skeleton of a river fish served as an allegory for the construct. Further, with a detailed study of the concept, it was decided to reduce the arch structures to the central semantic axis of the bridge. So, our arches pass the roadbed not along a parallel, but diagonally. This also creates a more dynamic composition and makes the design easier visually. Working with the roadways, we immediately came to the idea that we want to give most of the space for a public area, and not for transport. But we didn't forget about

the existing transport routes. Therefore, taking into account the closeness of the territory of Serebryany Bor for personal transport, we decided to organize two-lane traffic only for public transport (electric buses). So, the width of two lanes for electric buses is 7m; a bicycle path is 2.5m; a sidewalk with landscaping zones is 9m.

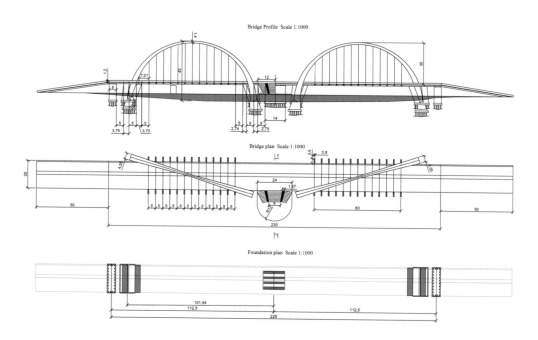

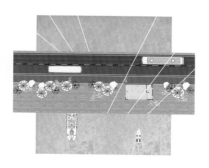

Concept collage

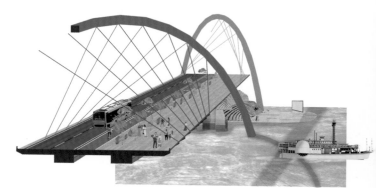

3D concept model collage

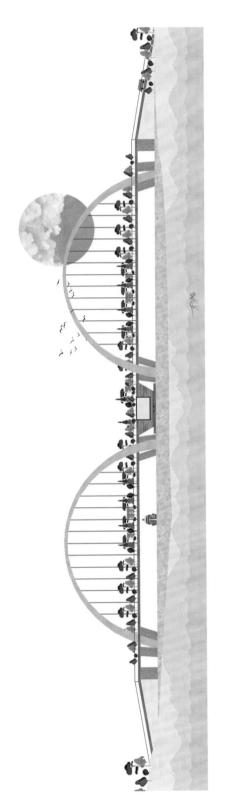

Plan collage

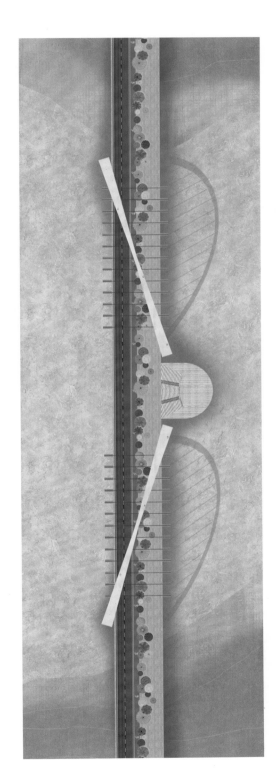

Profile collage

蝶梦桥

Butterfly Dream Bridge

学校名称 University/College Name

中国·沈阳建筑大学
Shenyang Jianzhu University, China

指导教师 Supervisor (s)

王占飞 WANG Zhanfei　　　　　宋福春 SONG Fuchun

参赛学生 Participant (s)

吴佳新 WU Jiaxin　　　　　苏洪业 SU Hongye
李明阳 LI Mingyang　　　　刘　帅 LIU Shuai
杨东方 YANG Dongfang

简介 Description

"庄周梦蝶"——蝶梦桥，俯视宛如敛翅蝴蝶，栩栩如生；正视宛如展翅蝴蝶，翩翩起舞，梦中蝴蝶寓意着人民对美好的物质生活和精神生活的追求，蕴含着我们每个人心中对未来的美好憧憬。

"Zhuang Zhou's Dream of Butterfly"—When looked down, the Butterfly Dream Bridge resembles a butterfly with folded wings. When faced upright, the bridge looks like a butterfly spreading its wings to dance lightly. The butterfly in the dream of Zhuang Zhou symbolizes people's pursuit of a better material life and spiritual life and their yearning for a brilliant future.

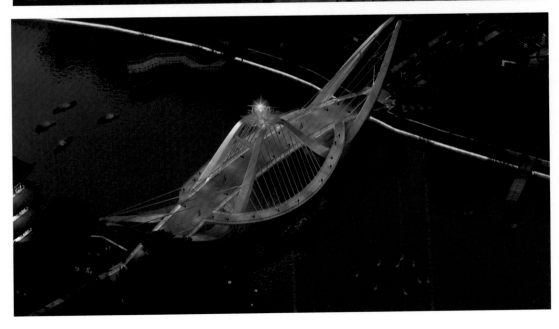

明月桥

Bright Moon Bridge

学校名称 University/College Name

中国 · 北京建筑大学
Beijing University of Civil Engineering and Architecture, China

指导教师 Supervisor (s)

焦驰宇 JIAO Chiyu

参赛学生 Participant (s)

朱　彬 ZHU Bin　　　　　　王寒冰 WANG Hanbing

李汉升 LI Hansheng　　　　刘文宇 LIU Wenyu

李嘉玺 LI Jiaxi

简介 Description

明月桥为中承式拱桥，总体设计采用饱含古诗韵味的月亮作为主要元素。结构上采用拱券外倾这一方案，取消了两拱肋之间的横撑，拱肋靠自身刚度和斜吊杆实现平衡，形似一轮新月，副拱券外倾，堪比海浪；桥梁两侧的副拱券相对于主梁方向外倾，形似飞溅而出的水花，整体看来仿若"海上生明月"。

项目设计吸收了西方近几年城市设计中的拱桥风格，注入了"中国水墨画"形神兼备的创意，并强调中华文化"天人合一"的艺术之魂。

The Bright Moon Bridge is a half-through arch bridge. The main element of the overall design is the moon, which is full of the charm of ancient poetry.

Structurally, the bridge applies an inclined arch ring and cancels the cross brace between the two arch ribs, relying on its own stiffness and balance of the inclined suspender. The bridge looks like a crescent moon, with the auxiliary arch ring tilted outward, which is similar to the sea wave. The secondary arch rings on both sides of the bridge incline outward relative to the direction of the main beam, looking like splashing water. The whole shape looks like the moon rising out of the sea.

In the project design, it absorbs the arch bridge of western urban design in recent years, injects the creativity of the "Chinese ink-wash painting" which stresses the unity of form and spirit, and emphasizes the artistic soul of "harmony between man and nature" of Chinese culture.

"云泽万邻"桥

Yunze Wanlin Bridge

学校名称 University/College Name

中国·沈阳建筑大学

Shenyang Jianzhu University, China

指导教师 Supervisor (s)

张怀志 ZHANG Huaizhi　　　　谢志伟 XIE Zhiwei

参赛学生 Participant (s)

王　谢 WANG Xie　　　　　　邢润强 XING Runqiang

肖清洋 XIAO Qingyang　　　　徐景余 XU Jingyu

陆彦羽 LU Yanyu

简介 Description

"一带一路"的国家开放战略与云南关系密切，云南是国家向西开放战略中的前沿和窗口，在整个宏观地理中具有突出的地位。而该桥正是修建于"一带一路"经济带中的云南楚雄地区两山之间，桥的跨径 200m，长度为 200m；桥面采用钢木叠合连续梁，宽度 11m，厚度 0.7m；主拱高度 16.5m，同为钢木组合结构，截面尺寸 1.65m×1.25m；主拱左右两边各设 2 个钢制桥塔柱以及 8 个木制小拱支撑桥面，小拱截面直径 1m；塔柱截面为 0.7m×0.7m。为了防雨，桥上面搭载涂满防腐蚀油漆的木制小拱防护篷，采用圆弧形设计对雨水进行导引和排放。桥下支撑和连接采用榫卯方式拼接，为减少风荷载，桥板中间结构采用波浪板设计。

The national opening-up strategy of the Belt and Road Initiative is closely related to Yunnan, the frontier and window of the country's westward opening-up strategy with a prominent position in the overall macro-geography. Therefore, the bridge is located in the Chuxiong region of Yunnan in the

Belt and Road Economic Belt. Built between the mountains, the bridge spans 200m with a length of 200m; the bridge deck is made of steel-wood laminated continuous beams, with a width of 11 m and a thickness of 0.7m; the main arch height is 16.5m, both of which are steel-wood composite structures, with a cross-sectional size of 1.65 m×1.25 m; on the left and right sides of the main arch, there are 2 steel bridge pylons and 8 small wooden arches to support the bridge deck. The diameter of the small arch is 1m; the cross section of the pylons is 0.7m×0.7m. To make it rainproof, the bridge is equipped with a protective canopy coated with anti-corrosion paint and supported by small wood arches, which adopts the arc-shaped design to guide and discharge rainwater. The supports and links underneath the bridge are spliced by mortise and tenon joints. In order to reduce wind load, the middle structure of the bridge slab adopts the design of the wave board.

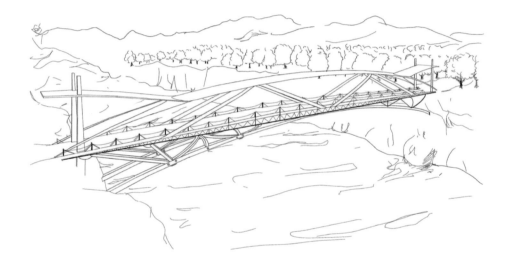

摄政运河人行天桥

Pedestrian Bridge over Regent's Canal

学校名称 University/College Name

英国·东伦敦大学
University of East London, the UK

指导教师 Supervisor (s)

Ali Abbas   Arya Assadi Langroudi

参赛学生 Participant (s)

Asif Kabir   Hanna Elizabeth Buckingham

Renver John Manuel Pena   Santhanalakshmi Sundararajan

Subash Chandra Limbu

简介 Description

　　煤场及国王十字街区位于伦敦卡姆登自治市，是全国各地货物储存的国家级枢纽。在 19 世纪，摄政运河主要用于将货物运往全国各地，然而，随着工业革命的兴起，水运逐渐衰败。本项目受起重机的启发，联想到 20 世纪皇家码头区内的货物贸易活动。桥梁的设计方案旨在为未来树立典范，并为拥有丰富工业历史和自然美景的该地区带来现代感。为了兼顾当地的繁荣发展和环境的宁静，这一方案包含斜拉和桁架桥设计两部分，并提供了一种有效的悬臂解决方案，通过斜拉索承受来自桥面的荷载，从而避免其他方案可能对环境造成的干扰。

Being located in the London borough of Camden, the Coal Drops Yard and King's Cross Areas served as a national hub for the storage of goods from all over the country. The Regent's Canal was mainly serued as a mean of transporting goods all over the country during the 19th century, however, with the rise of the industrial revolution, water transport became obsolete. A crane, which serves as inspiration, harkens back to the movement of goods and trade within the Royal docks area during the 20th century. The bridge design scheme aims to become a monument of the future, bringing a modern twist to an area with a rich industrial history as well as natural beauty. In the effort to link the boom of local area development with the serenity of the surrounding nature, the design composed of a combination of cable stay and truss bridge design schemes. It is an effective cantilever solution to bear the loads from the bridge deck through cable stays, preventing detrimental disturbances to the beauty of the surrounding environment which alternative schemes may further contribute to.

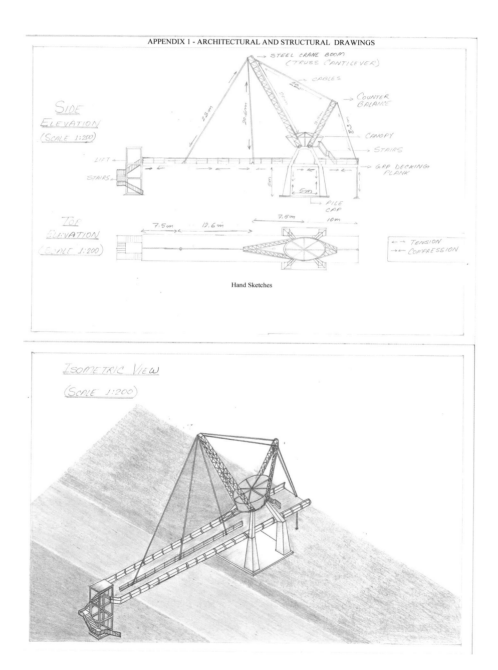

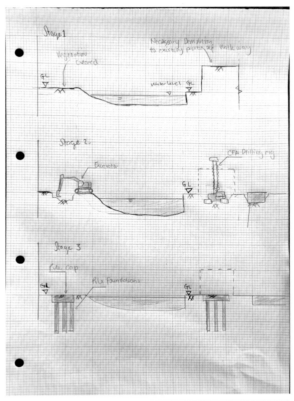

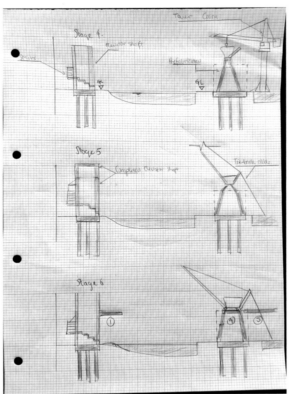

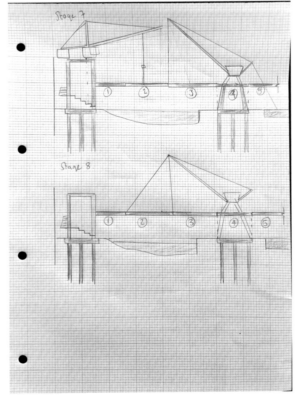

东岳桥

Dongyue Bridge

学校名称 University/College Name

中国·西安建筑科技大学
Xi' an University of Architecture and Technology, China

指导教师 Supervisor (s)

孙建鹏 SUN Jianpeng

参赛学生 Participant (s)

谭子涵 TAN Zihan   主父高林 ZHUFU Gaolin

王忠康 WANG Zhongkang   冯怡杰 FENG Yijie

Saitoti Emanuel Elibariki

简介 Description

    东岳桥连接对岸与小九华寺，位于苏州市吴江区平望镇小九华路 1 号，横跨京杭大运河，桥位所处运河宽度为 35m 左右。该桥连接的小九华寺建于明万历四十四年，原名东岳庙，祀东岳泰山之神。我们将桥梁取名为东岳桥，也是为了纪念寺庙的旧名。桥型选为拱桥，寓意岳字，主要作为人行景观桥。本桥横跨京杭大运河，结合小九华寺这一历史古迹和风景名胜莺湖，有利于中国一带一路"大运河文化带"的构建，与本次赛事主题不谋而合。

Dongyue Bridge connects the opposite side with Xiaojiuhua Temple, located at No.1 Xiaojiuhua Road, Pingwang Town, Wujiang District, Suzhou City, crossing the Beijing-Hangzhou Grand Canal. The bridge is located where the canal width is about 35m, connecting Xiaojiuhua Temple, which was built in the 44th year of the Ming Dynasty as Dongyue Temple, offering the god of Mount Taishan. We named the bridge Dongyue Bridge also as a way to commemorate the old name of the temple.

The bridge type is selected as an arch bridge, symbolizing the Chinese character "Yue", mainly functioning as a pedestrian landscape bridge. Spanning the Beijing-Hangzhou Grand Canal, and together with the historical site Xiaojiuhua Temple and the scenic spot Yinghu Lake, the bridge is conducive to the construction of the "Grand Canal Cultural Belt" in the Belt and Road Initiative, which is in line with the theme of this competition.

红旗大桥

Hongqi Bridge

学校名称 University/College Name

中国·河南城建学院

Henan University of Urban Construction, China

指导教师 Supervisor (s)

屈讼昭 QU Songzhao   董召锋 DONG Zhaofeng

参赛学生 Participant (s)

侯澳国 HOU Aoguo   张家诚 ZHANG Jiacheng

张新亮 ZHANG Xinliang   曾兴涛 ZENG Xingtao

王占岭 WANG Zhanling

简介 Description

本设计为一预应力混凝土箱式截面的 T 型刚构桥设计，桥长为 50m，跨径组合 (2×25)m，桥面宽为 5.5m=0.5m（防撞护栏）+4m（行车道）+0.5m（防撞护栏），汽车荷载等级：公路 II 级。

桥梁采用预应力混凝土变高度箱型梁（A 类构件），采用后张法工艺进行上部主梁的施工，预应力钢筋采用 1860 钢绞线，HVM15 型锚具，C50 混凝土，HRB400 普通钢筋，施工方法选用满樘支架施工，轻型桥台，薄壁式桥墩，并采用矩形扩大基础。

This design is a T type rigid frame bridge design of prestressed concrete box section, the bridge length is 50 m, the span combination uses (2×25) m, the width of the bridge is 5.5 m=0.5 m (anti-collision guardrail) +4 m (lane) +0.5 m (anti-collision guardrail), the carload level: highway-II level.

The bridge applies prestressed concrete box beam consisted of variable height (class A) components, using the post-tensioned method process for the upper main girder construction, prestressed reinforced by 1860 strands. The anchorage is HVM15 type with C50 concrete and HRB400 ordinary steel. The construction method is full support construction, using a light abutment, thin-walled bridge piers, and a rectangle enlarged foundation.

可居住桥梁多功能文化中心项目

Project of a Multifunctional Cultural Center on an Inhabited Bridge

学校名称 University/College Name

俄罗斯·圣彼得堡国立建筑大学

St. Petersburg State University of Architecture and Civil Engineering, Russia

指导教师 Supervisor (s)

Kolodin Konstantin

参赛学生 Participant (s)

Tretyakova Polina

简介 Description

建筑与工程结构共生于可居住桥梁。除主要的交通和步行功能外，可居住桥梁还具有住宅或公共建筑的功能。圣彼得堡拥有约 800 座桥梁，共 93 条河流和运河。桥梁是这座城市的标志，包括吊桥、斜拉桥、拱桥和梁桥，但却没有可居住桥梁。在我看来，建造可居住桥梁这一主题与我们的城市息息相关，此举将创造一种新型结构和城市新标志。因此，我提议在一座可居住桥梁上设立一个文化中心。该项目旨在满足在休闲区、商业开发区和住宅区间修建步行通道的需求。

Habitable bridge is a symbiosis of architecture and engineering structure. These are objects that, in addition to their main transport and pedestrian functions, performing the function of a residential or public building. Saint-Petersburg is the city, in which about 800 bridges, 93 rivers and canals in total. Bridges here are the hallmark of the city, such as drawbridges, cable-stayed bridges, arched bridges and beam bridges. But there is not a single habitable bridge. In my opinion, the theme of building a such bridge is relevant for our city. This will create a new type of structures and a new symbol of the city. Therefore, I propose to create a cultural center on a habitable bridge. This project is

2021 国际大学生建筑设计与数字建模竞赛作品集
2021 International Student Competition on Architectural Design and Digital Modelling Work Collection

due to the need for a pedestrian connection between the recreational area, business development and residential area.

preliminary possible design schemes

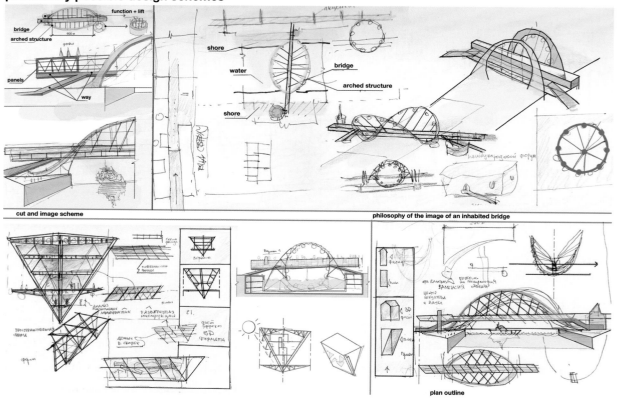

青芙绸桥

Qingfu Silk Bridge

学校名称 University/College Name

中国·长春建筑学院
Changchun University of Architecture and Civil Engineering, China

指导教师 Supervisor (s)

李厚萱 LI Houxuan　　　　　　张晓娇 ZHANG Xiaojiao

参赛学生 Participant (s)

卢秋瑶 LU Qiuyao　　　　　　熊世奇 XIONG Shiqi

寻之冶 XUN Zhiye　　　　　　刘　奇 LIU Qi

刘峻宇 LIU Junyu

简介 Description

相对于其他桥梁结构，青芙绸桥使用较少的材料跨越了较长的距离。而对于传统桥梁来说，跨越较长的距离需要大量的材料来进行加固，费时、费财、费力。这不利于对跨度较大的江河上的施工。青芙绸桥可以造得比较高，容许船在下面通过，在造桥时没有必要在桥中心建立临时的桥墩，因此可以在这种比较深或比较急的水流上建造悬索桥。

Compared with other bridge structures, Qingfu Silk Bridge uses fewer materials to cover a longer distance. For the traditional bridge, to span a long distance requires a lot of reinforcing materials, which consumes time, finance, and labor. It doesn't fit wide rivers. The bridge can be built high enough to allow ships to pass below. It is not necessary to build temporary piers in the center of the bridge, so suspension bridges can be built over deeper or more rapid currents.

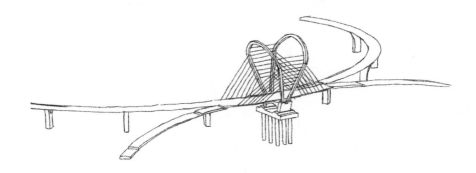

东栅·雨眠
East Gate & Rainy Night

学校名称 University/College Name

中国·长春建筑学院
Changchun University of Architecture and civil Engineering, China

指导教师 Supervisor (s)

尹鹏程 YIN Pengcheng   王星博 WANG Xingbo

参赛学生 Participant (s)

杨泺怡 YANG Luoyi   芦 伟 LU Wei

秦鸿艳 QIN Hongyan   田驭洲 TIAN Yuzhou

闫威峰 YAN Weifeng

简介 Description

古往今来，一提起值得一去的美景，肯定少不了江南水乡。小桥、流水、人家，所到之处尽是闲适舒缓之景，时光静谧，着实令人心旷神怡。乌镇东栅完美地还原了所需要的全部要点，一场乌镇之游一定会让你心旷神怡。

本次设计先运用 CAD 确定测量数据以及局部细节，之后再以数字建模的方式来一比一还原东栅的各种景色以及当地丰富的乡土人情，最后渲染出图。目的是让学生发挥主观能动性，动手收集材料，和不同专业的伙伴相互配合，也在制作过程中体验中国江南的水乡文化。

Throughout the ages, when it comes to the beauty that is worth visiting, Jiangnan Water Village is definitely not to be missed. Small bridges, flowing waters, cottages, wherever you go, there is relaxing and soothing scenery, which is really tranquilizing and refreshing. Wuzhen's East Gate perfectly restores all the essentials needed, and a trip to Wuzhen will definitely make you feel relaxed and happy.

The design starts with CAD, measuring data and local details required, then uses digital modeling to restore the various scenery of the East Gate and the rich local culture before the final rendering. We hope that students can exercise their own initiative to collect materials, cooperate with different majors, and experience China's Jiangnan Water Village culture during the production process.

一带一路：梦忆南阳

Belt and Road: Dreaming of Nanyang Ancient Town

学校名称 University/College Name

中国·北京建筑大学
Beijing University of Civil Engineering and Architecture, China

指导教师 Supervisor (s)

廖维张 LIAO Weizhang   刘 扬 LIU Yang

参赛学生 Participant (s)

高思岩 GAO Siyan   刘若清钰 LIU Ruoqingyu

阚效禹 KAN Xiaoyu   孙 尹 SUN Yin

刘泽华 LIU Zehua

简介 Description

历史悠久的运河文化，南阳古镇是微山湖中运河线上最具有特色的城镇，面积约 4.5km²，其古代建筑群是中华文化的重要传承之一，是值得后人保护和学习的。以康乾别院文物古迹、状元楼和魁星楼建筑物为主，植被、运河和居民建筑等自然景观为辅，现场采集无人机倾斜摄影测量与三维激光扫描数据，并使用 ContextCpature 和 Navisworks 内业处理软件进行建模，实现南阳古镇建筑群与场景数字建模，旨在认识传统建筑文化的优秀基因，传承与发扬古建筑模型的制作工艺。

With a long history of canal culture, Nanyang Ancient Town is the most distinctive town on the canal line in Weishan Lake, with an area of about 4.5km2. Its ancient complex is one of the important inheritances of Chinese culture which is worthy of protection and learning by future generations. This experiment focuses on the cultural relics and historic sites of the Kang-Qian Courtyard, the Zhuangyuan Building and the Kuixing Tower, supplemented by natural landscapes such as

2021 国际大学生建筑设计与数字建模竞赛作品集
2021 International Student Competition on Architectural Design and Digital Modelling Work Collection

vegetation, canals and residential buildings. We collected oblique photogrammetry and 3D laser scanning data on site, and used ContextCpature and Navisworks office processing software for the digital modeling of buildings and scenes in the Nanyang Ancient Town, aiming to identify the

excellent genes of traditional architectural culture, and to inherit and carry forward the production

process of ancient architectural models.

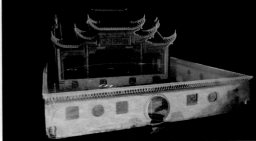

"塔影遥瞻碧水隈，梵音远传广陵乡"

"The Pagoda Overlooks Its Shadow on the Clear Water, and the Sound of the Buddhist Spreads As Far As the Township of Guangling"

学校名称 University/College Name

中国 · 北京建筑大学

Beijing University of Civil Engineering and Architecture, China

指导教师 Supervisor (s)

黄　鹤　HUANG He　　　　冯永龙　FENG Yonglong

参赛学生 Participant (s)

曹聿铭　CAO Yuming　　　　刘羿婷　LIU Yiting

赵嘉仁　ZHAO Jiaren　　　　朱欣宇　ZHU Xinyu

顾泊桐　GU Botong

简介 Description

利用无人机进行倾斜摄影后使用大疆智图建模并导出模型文件，将导出的模型文件加载到电脑中使用 3D MAX 进行后期的精细化处理。用 3D MAX 生成 obj 格式的三维数据，找出需要修改和校正的三维瓦片，最后，选中的瓦片通过 3D MAX 软件的线面编辑工具得到修补。处理之后修整水面，选择模型文件之后在修改器列表中找到噪波修改器。修改强度下的参数，修改比例值调整波长，调整种子的值以获得不同效果，调整分形设置粗糙度，最后更改水面颜色并对破损的建筑物补洞后模型基本完成。

The drone was used for tilt photography to export the model file by DJI Wisdom. The exported model file was loaded into the computer using 3D MAX for post refinement. 3D MAX was used to generate 3D data in obj format, to find out the 3D tiles that needed to be repaired and modified.

4

Finally, the selected tiles were repaired by the 3D MAX software's line surface editing tool for the missing model. After processing to fix the water surface, the model file was selected and the noise waver modifier was found in the list of modifiers. The parameters under intensity were modified, the scale value was modified, and the wavelength was adjusted. The value of the seed was adjusted to get different effects, and the fractal setting was adjusted to set the roughness. Finally, the color of the water surface was changed and the hole of the broken building was patched before the model was basically formed.

建筑与场景 3D 数字建模方向
Category C: Building and Space 3D Digital Modeling

蜂　旅

Bee Hotel

学校名称 **University/College Name**	中国·山东建筑大学 Shandong Jianzhu University, China
指导教师 **Supervisor** (s)	张莉莉 ZHANG Lili
参赛学生 **Participant** (s)	王丽婷 WANG Liting

简介 Description

蜂旅被构想为一座主题度假休闲地，不仅具备民宿功能，还有主题风格的小店、闲适的茶亭，拥有朝向前运河和后方林木的双向视野，完全可以供人们在这里享受温暖且充满户外气息的假期生活。

建筑形式上以蜂巢结构为参考，蜂巢内外面的巢穴刚好两两相互错开，相互组合六角形的边交叉的点是内侧六角形的中心，提高了蜂巢的强度，防止巢房底破裂。蜂旅主体建筑便是以独立的六边体单体空间构成一个个独立或可连接的蜂巢状建筑。

Bee Hotel is conceived as a thematic vacation and leisure place. It not only has the function of B&B, but also thematic small shops and civilized tea pavilion. It boasts a two-way view of the forward canal and the rear trees, which can fully provide people with a warm outdoor holiday life.

The architectural form takes the honeycomb structure as the reference. Half of the nest inside and outside the honeycomb stagger each other, with the six sides of the hexagon intersecting at the center of the inner hexagon, which reinforces the intensity of the honeycomb and prevents the bottom from breaking. The body of the Bee Hotel comprises separate or connected honeycomb buildings with a separate hexagonal monomer space.

归 源

Back to the Source

学校名称 University/College Name

中国·吉林建筑大学
Jilin Jianzhu University, China

指导教师 Supervisor (s)

宋义坤 SONG Yikun　　　　　　安　宁 AN Ning

参赛学生 Participant (s)

谷元振 GU Yuanzhen　　　　　温　静 WEN Jing

冯　玥 FENG Yue　　　　　　范馨穗 FAN Xinsui

左一轩 ZUO Yixuan

简介 Description

数百年来，明朝古楼以各种不同的角色见证着运河的兴衰往事，无论是文化象征还是战略需求，每一种存在都是运河发展历史中的重要印记。本项目旨在发掘古楼的历史记忆，重拾新时代的古楼价值。

For hundreds of years, the ancient buildings of the Ming Dynasty have witnessed the rise and fall of the canal with various roles. Whether it is a cultural symbol or a strategic need, each form is an important mark in the history of the canal's development. This project aims to discover the historical memory of ancient buildings and regain the value of ancient buildings in the new era.

桥衔古今

The Bridge of Memory

学校名称 University/College Name

中国·吉林建筑大学

Jilin Jianzhu University, China

指导教师 Supervisor (s)

宋义坤 SONG Yikun   赵建彤 ZHAO Jiantong

参赛学生 Participant (s)

戴沈周 DAI Shenzhou   孙文昊 SUN Wenhao

吴哲涵 WU Zhehan   郭 璐 GUO Lu

陈嘉怡 CHEN Jiayi

简介 Description

桥是京杭大运河交通系统的重要组成部分，也是古往今来人们的生产、生活与运河水系发生关联的重要载体和纽带。在蜿蜒数千里的运河上无数的古桥承载着人们对美好生活的渴望、对沧桑历史的慨叹以及船夫们的乡愁。本建模设计选取京杭大运河江南段一处历史桥梁为对象，旨在对其进行记录和保护。

Bridges are important parts of the transportation system of the Beijing-Hangzhou Grand Canal, as well as an important carrier and link between people's production, life and the canal system throughout the ages. The numerous ancient Bridges along the canal winding for thousands of miles carry people's longing for a better life, their lament for the vicissitudes of history and the boatmen's homesickness. This modeling design selects a historical bridge in the south section of the Beijing-Hangzhou Grand Canal, aiming to record and protect it.

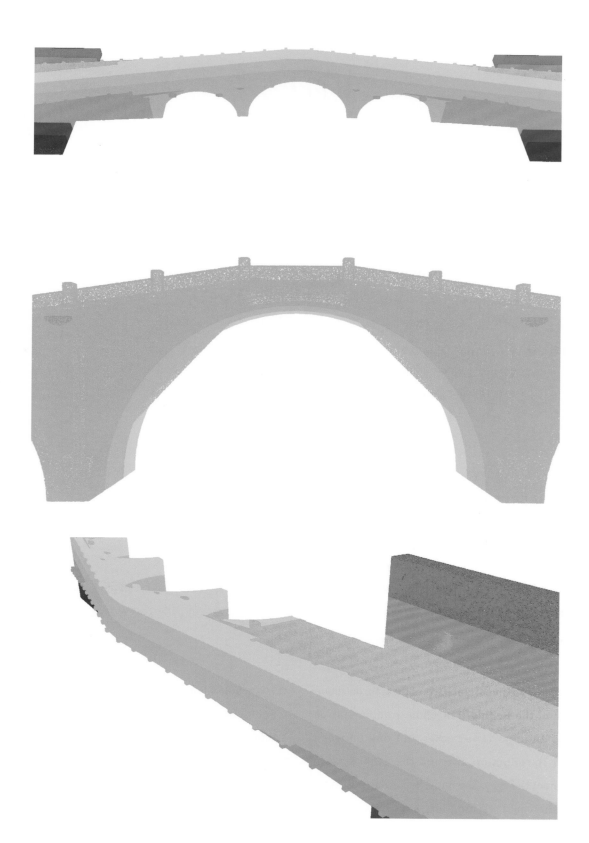

盘云水和

Harmony around Everything

学校名称 University/College Name
中国·山东建筑大学
Shandong Jianzhu University, China

指导教师 Supervisor (s)
张莉莉 ZHANG Lili

参赛学生 Participant (s)
王艺霏 WANG Yifei　　　　　　　王一诺 WANG Yinuo

简介 Description

项目所选古城位于京杭大运河沿岸的山东省枣庄市辖区台儿庄区，地处山东省的最南端。既是民族精神的象征、历史的丰碑，也是运河文化的承载体，被誉为"活着的运河""京杭大运河仅存的遗产村庄"。重建台儿庄古城，是几代台儿庄人民的梦想。

The project selected the ancient city along the Beijing-Hangzhou Grand Canal in Taierzhuang District, Zaozhuang City, Shandong Province. The ancient city lies on the southernmost point of Shandong Province. Known as the "living canal" and "the only remaining heritage village along the Beijing-Hangzhou Canal", it is not only a symbol of national spirit, a historical monument, but also a carrier of canal culture. For several generations of Taierzhuang people, they dream of rebuilding the ancient city of Taierzhuang.

学古通今，运河神韵——运河岸学堂设计

Learn from Ancient Times to the Present, the Charm of the Canal —Design of the School to the Canal Bank

学校名称 University/College Name

中国・天津交通职业学院

Tianjin Transportation Technical College, China

指导教师 Supervisor (s)

任杨茹 REN Yangru   迟文伟 CHI Wenwei

参赛学生 Participant (s)

王　义 WANG Yi   李河川 LI Hechuan

王　菲 WANG Fei   张　昭 ZHANG Zhao

杜德轩 DU Dexuan

简介 Description

　　本作品采用 BIM 建模，建筑基本还原运河的神韵，杨柳青镇隶属于天津市西青区，系西青区政府驻地，是西青区的政治、经济和文化中心。地处京畿要冲，位于津城西厢，东临中北镇、西营门街道，西有张家窝镇，西南接辛口镇，西北连武清区，北临北辰区。2017 年，总面积 6517 公顷，总人口161247 人，是天津市与环渤海经济区最大的乡镇。杨柳青镇历史沉淀久远，文化底蕴深厚。杨柳青镇有丰富民间艺术。2018 年 5 月 24 日，杨柳青镇入选最美特色小城镇 50 强。2019 年 1 月 9 日，杨柳青镇凭借杨柳青木版年画入选 2018—2020 年度"中国民间文化艺术之乡"名单。 2018 年再次荣登国家卫生县城（乡镇）之列。

This work adopts BIM modeling, and the buildings basically restore the charm of the canal.

Located in Xiqing District of Tianjin Municipality, Yangliuqing Town is the resident of Xiqing District

People's Government of Tianjin Municipality, and the political, economic and cultural center. In terms of geographical location, the town plays a critical role in safeguarding the capital. To be more specific, Yangliuqing Town lies on the western part of Tianjin, bordering Zhongbei Town and Xiyingmen Street in the east, Zhangjiawo Town in the west, Xinkou Town in the southwest, Wuqing District in the northwest, and Beichen District in the north. In 2017, with a total area of 6,517 hectares and a total population of 161,247, it is the largest town in Tianjin and the Bohai Rim Economic Zone. The time-honored Yangliuqing Town boasts profound cultural deposits and rich folk art. On May 24, 2018, it was selected as the top 50 beautiful characteristic small towns. On January 9, 2019, the town was included in the list of Chinese Folk Culture and Art Townships in 2018-2020 because of its New Year woodblock prints. In 2018, it was reconfirmed as a National Health County (Township).

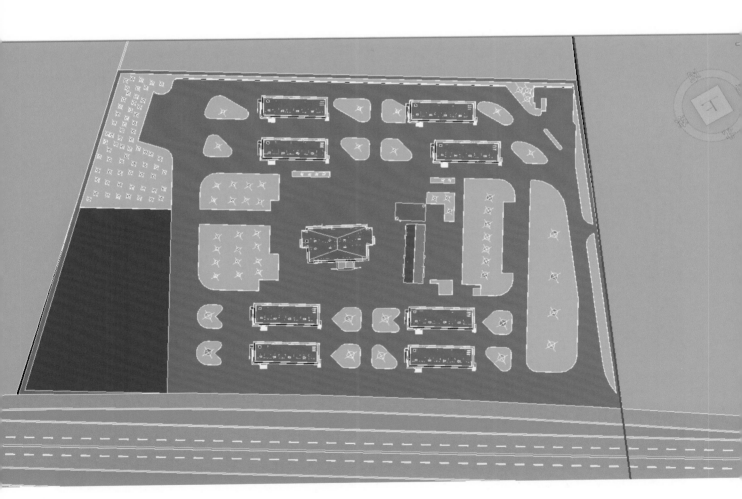

2021 国际大学生建筑设计与数字建模竞赛作品集
2021 International Student Competition on Architectural Design and Digital Modelling Work Collection

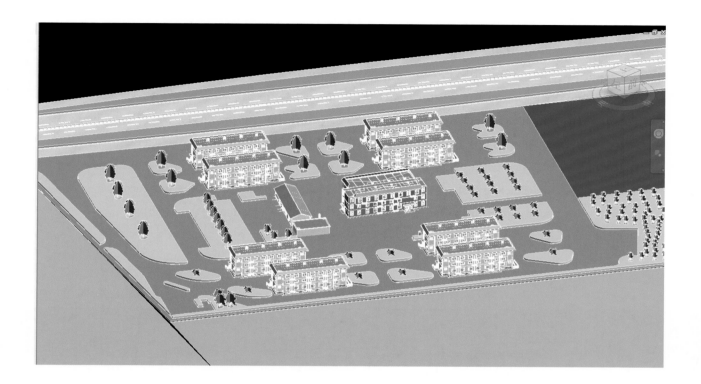

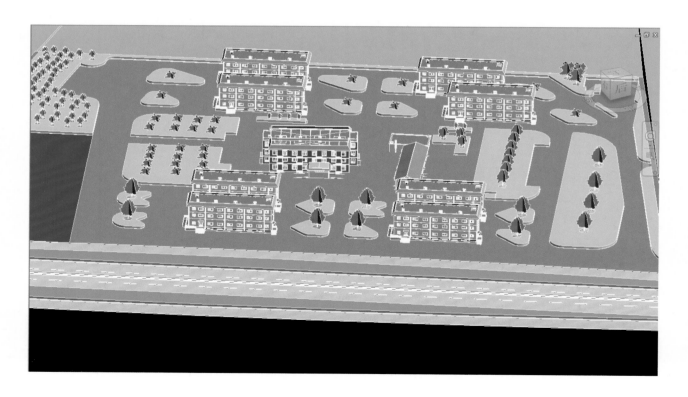

雁鸣码头

Yanming Wharf

学校名称 University/College Name

中国·西安理工大学
Xi'an University of Technology, China

指导教师 Supervisor (s)

田小红 TIAN Xiaohong

参赛学生 Participant (s)

方　虹 FANG Hong
李小英 LI Xiaoying

汤艺琳 TANG Yilin

简介 Description

　　茶馆的装修设计，主在空间气氛的营造。设计师在此房间的中式设计说明中阐释：浓郁的茶香和禅意，隐而不露的暗格雕花，凸显中式的皇室金龙壁画、宫灯吊饰……通过诸多设计元素的合理排布，将禅意和茶道、古韵和中式感觉结合融合一体，凝结为该茶馆集中式百家的大成之作。空间中严肃而不失温润。中式设计说明了中式文化在当代，一样会受到人们的接纳和赞许。

Teahouse decoration design focuses on the spatial atmosphere. In the design, the room, the rich tea fragrance, the Zen, the hidden pattern carvings, the Chinese royal golden dragon fresco, the palace lantern hanging ornaments, etc., are reasonably arranged to integrate the Zen, tea ceremony and traditional Chinese elements to render the design the pastiche of extensive Chinese styles. The space is solemn yet gentle. The Chinese style of design shows how Chinese culture can be accepted and praised in contemporary times.

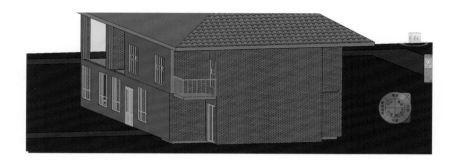

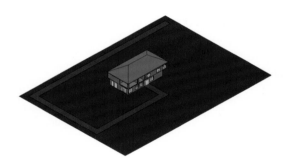

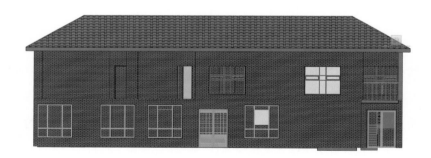

通州燃灯佛舍利塔

Tongzhou Dipamkar Stupa

学校名称 University/College Name

中国·北京建筑大学
Beijing University of Civil Engineering and Architecture, China

指导教师 Supervisor (s)

王国利 WANG Guoli   郭　明 GUO Ming

参赛学生 Participant (s)

付泽昕 FU Zexin   梁轩硕 LIANG Xuanshuo

张　艾 ZHANG Ai   甄建超 ZHEN Jianchao

周玉泉 ZHOU Yuquan

简介 Description

　　燃灯佛舍利塔位于北京市通州区，北运河源头岸边，是北京市重点保护文物。本设计对舍利塔及其周边环境进行了全面数字化数据采集及建模分析工作：用三维激光扫描与低空无人机摄影技术对塔进行了毫米级精细三维几何与纹理数据采集，以采集数据为基础，利用三维建模软件 PhotoScan、SketchUp、3D MAX 对不同对象进行了分类建模并完成模型集成，利用获取的精密三维数据对塔的歪闪、扭曲等健康现状进行了全面分析，充分展现了塔的健康现状及环境状态。

　　The Depamkar Stupa is located in the Tongzhou District of Beijing, at the source of the northern section of the Grand Canal. It is a key cultural relic under the protection of Beijing Municipal Government. The design carries out comprehensive digital data collection and modeling analysis of the stupa and its surrounding environment. Millimeter-level fine 3D geometry and texture data of the stupa are collected with 3D laser scanning and low altitude UAV photography technology. Based on the data,

2021 国际大学生建筑设计与数字建模竞赛作品集
2021 International Student Competition on Architectural Design and Digital Modelling Work Collection

different objects are classified, modeled and integrated by using 3D modeling software such as PhotoScan, SketchUp and 3D MAX. The health status of the stupa, such as slanting and distortion, is comprehensively analyzed with precise 3D data, which shows the present condition and environment of the stupa.

通州燃灯佛舍利塔

　　燃灯佛舍利塔，矗立于京杭大运河北端，是中国运河四大名塔之一，为运河岸边标志性建筑，燃灯佛舍利塔是最古老的。始建于南北朝北周宇文时期，距今已有1300多年历史，已被列为北京市重点保护文物。该塔为砖木结构，密檐实心，共八角十三层，高约49m，塔基须弥座呈莲花形。燃灯佛舍利塔与佑胜教寺相邻，成为京杭大运河北端之胜景，也是通州地区具有代表性的建筑遗产之一。但是目前有关燃灯佛舍利塔的相关信息多为纸质资料及文字档案，三维数字化资料及详细结构信息严重缺乏，因此对塔进行三维数字化测量变得尤为重要。

信息采集（三维激光扫描、无人机拍摄）

Z+F5010扫描仪　　Rigle VZ1000扫描仪

三维激光扫描测站分布

　　项目采用远程与近程两种地面三维激光扫描仪，通过对现场及周围环境勘察，合理布置扫描点，保证获得完整的扫描数据。点云数据采集按照7mm表面分辨率进行采集。其中Z+F5010扫描仪设有8个测站，扫描点近距离（约50m）精度可达1~2mm。Rigle VZ1000扫描仪设有6个测站，一次单点扫描精度为5mm（100m距离处）。

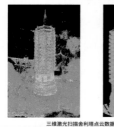 

三维激光扫描舍利塔点云数据

　　采用低空无人机拍摄，在塔的8个面分别沿直线自下而上拍摄。共获取约680张影像，其拍摄表面分辨率0.1mm/像素，单张影像像幅为5616像素×3744像素。

nikon单反相机

8悬翼无人机

无人机影像实例

燃灯佛舍利塔完整数据

三维建模

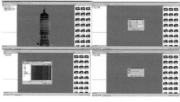 

photoscan数据处理过程

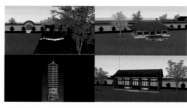 

建模过程局部图

建模成果图

病害分析

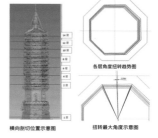

各层角度扭转趋势图

横向剖切位置示意图　　扭转最大角度示意图

纵向剖切位置示意图

　　对塔每层斗拱下部位置进行横向剖切，通过分析提取各层边界，得到二层与顶层相对扭转的最大角度为1.578°，其余各层扭转角度在此区间内随机分布。

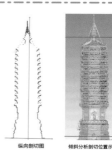

纵向剖切图

倾斜分析剖切位置示意图

倾斜分析图

　　通过数据处理得到燃灯佛舍利塔基座几何中心和塔尖部位的几何中心，将两个中心连线，分析可得，塔尖中心相对于基座中心当前偏移距离是0.3601m，当前角度偏移是西偏南41.3566°。

Standing at the northern end of the Beijing-Hangzhou Grand Canal, Depamkar pagoda is one of the four famous pagodas of the Chinese Canal and a landmark building along the canal bank. Depamkar pagoda is the oldest. Built in the Northern Zhou Yuwen Period of the Northern and Southern Dynasties, it has a history of more than 1000 years and has been listed as a key cultural relic under the protection of Beijing. The pagoda is a brick and wood structure, dense eaves solid, a total of eight corners, 13 stories, about 45 meters high, the base is lotus shaped. Depamkar pagoda, a scenic spot at the northern end of the Beijing-Hangzhou Grand Canal, is also one of the representative architectural heritages in Tongzhou area. However, at present, the relevant information about Depamkar pagoda is mostly paper data and text files, and 3D digital data and detailed structural information are seriously lacking. Therefore, it is particularly important to carry out 3D digital measurement on the pagoda.

Information acquisition （3D laser scanning、The UAV aerial photography）

ZZ+F5010 scanner Rigle VZ1000 scanner

Distribution of 3D laser scanning stations

The project adopts two kinds of ground 3D laser scanners, which are long distance and short range. Through the investigation of the site and surrounding environment, the scanning points are reasonably arranged to ensure the complete scanning data. Point cloud data acquisition was carried out according to the surface resolution of 7mm. The Z+F5010 scanner has 8 measuring stations, and the accuracy of the scanning point at a short distance (about 50m) can reach 1~2mm. Rigle VZ1000 scanner with 6 stations, a single point scanning accuracy of 5mm (100m distance).

Three-dimensional laser scanning of the point cloud data

the example of UAV image

Using low altitude UAV photography, the tower's eight faces were photographed along a straight line from bottom to top. A total of about 600 images were obtained, with a surface resolution of 0.1mm/pixel, and a single image size of 5615 pixels * 7144 pixels.

Nikon SLR camera

8 Suspended UAV

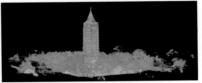
Complete data of Depamkar pagoda

3D modeling

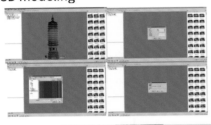

photoscan Data processing process

Local diagram of the modeling process

Modeling results diagram

Disease analysis

Schematic diagram of transverse section position

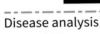
Angle torsion trend chart of each layer

Schematic diagram of maximum torsion Angle

Diagram of longitudinal cutting position

The lower part of each floor of the tower was cut horizontally, and the boundary of each floor was extracted through analysis. The maximum relative torsion angle between the second floor and the top floor was 1.378°, and the torsion angles of the other floors were distributed randomly in this interval.

Longitudinal section drawing

Diagram of cutting position for tilt analysis

Tilt analysis diagram

Through data processing, the geometric centers of the base and the spire of Depamkar pagoda were obtained. By connecting the two centers and analyzing, the current deviation distance between the center of the spire and the base center is 0.360 m, and the current angle deviation is 41.2566° by south of west.

水韵古城

Ancient City of Water

学校名称 University/College Name

中国·北京建筑大学

Beijing University of Civil Engineering and Architecture, China

指导教师 Supervisor (s)

董友强 DONG Youqiang   侯妙乐 HOU Miaole

参赛学生 Participant (s)

王　飞 WANG Fei   王　茜 WANG Qian

李博仑 LI Bolun   栗怡豪 LI Yihao

杨　玥 YANG Yue

简介 Description

大明湖集景观、园林、建筑、水域为一体，具有悠久的历史、深厚的文化积淀和高品位的旅游资源。利用无人机倾斜摄影测量技术和 ContextCapture 软件可以获得大明湖及园内建筑高精度的实景三维模型。还可以获取海量的地理信息数据。通过数字建模可以展现大明湖的真实场景，展现济南泉城文化，城市园林文化及老济南的民族文化，为后续景区的保护利用和文化的展示传播提供巨大帮助，彰显济南的区域文化特色，以数字建模技术见证、记录和促进文明互鉴，让中国文化远播重洋。

Daming Lake integrates landscape, garden, architecture and water area, and boasts a long history, profound cultural accumulation and high-standard tourism resources. The high-precision three-dimensional model of Daming Lake and the surrounding architecture can be obtained by using UAV oblique photogrammetry technology and ContextCapture software. A large amount of geographic information data can also be acquired. Through digital modeling, we can show the real scenery of Daming Lake, showcase the rich water resources of Jinan City, urban garden culture and the national culture exemplified in ancient Jinan, which is of great help for the protection and utilization of scenic spots and the communication of culture. The model can highlight the regional cultural characteristics of Jinan, and witness, record and promote mutual learning among civilizations with digital modeling technology, so that Chinese culture can spread across the globe.